Design for Structural Stability

The Constrado Monographs deal with the application of steel in construction. They each treat a specific subject, and the texts are written with authority and expedition. Subjects are treated in depth and are taken to the point of practical application.

ADVISORY EDITOR

M. R. Horne, MA, ScD, FICE, FIStructE
Professor of Civil Engineering, University of Manchester

ALREADY PUBLISHED

The Stressed Skin Design of Steel Buildings
by E. R. Bryan

Thin Plate Design for Transverse Loading
by B. Aalami and D. G. Williams

Composite Structures of Steel and Concrete
Volume 1
by R. P. Johnson
Volume 2
by R. P. Johnson and R. J. Buckby

Thin Plate Design for In-Plane Loading
by D. G. Williams and B. Aalami

CONSTRADO MONOGRAPHS

Design for Structural Stability

P. A. Kirby and D. A. Nethercot

A HALSTED PRESS BOOK

JOHN WILEY & SONS
New York

Published in Great Britain by Granada Publishing Limited
in Crosby Lockwood Staples 1979

Published in the U.S.A.
by Halsted Press, a Division of
John Wiley & Sons, Inc., New York

Library of Congress Cataloging in Publication Data
Kirby, P. A.
   Design for structural stability.

   (Constrado monographs)
   "A Halstead Press book."
   Bibliography: p.
   1.   Structural stability.   2.   Building, Iron and
steel.   I.   Nethercot, D. A., joint author.
II.   Title.
TA656.K57   1979   624'.1821   79–754
ISBN 0–470–26691–0

Printed in Great Britain by Richard Clay (The Chaucer Press) Ltd,
Bungay, Suffolk

Contents

Foreword

The requirements of structures in relation to strength and stiffness are largely amenable to intuitive understanding. This means that, although extensive calculations may be involved in designing and checking structures to satisfy, for example, requirements of allowable stresses or permissible deflections, engineers easily absorb the significance of the quantities they are calculating, and have a ready appreciation of their physical implication. This very much lessens the likelihood of errors, should they arise, remaining unnoticed. The same cannot be said of the structural phenomenon of instability, whereby a member of apparently sufficient strength to withstand the forces to which it is subjected, somehow buckles as it were 'out of the way' of the applied forces, leading to catastrophic failure. Witness to the lack of intuitive understanding of structural stability by engineers is provided by many failures throughout the history of structural engineering, in which stability requirements have been seriously underestimated.

The principles of structural stability cannot be understood without some willingness to master the mathematics of linear differential equations. Although all engineers deal with these well enough in their undergraduate courses, there is a tendency amongst not a few practising engineers to regard engineering concepts involving anything beyond O-level mathematics with distinct suspicion. Certainly, texts on elastic stability — many excellently written, in the style pioneered by Timoshenko — are commonly regarded as 'academic' and beyond the understanding of the average engineer. To the extent that this attitude persists, it is regrettable, and goes some way to explain the frequency of failure attributable to some form of structural instability.

In the new generation of codes based on ultimate capacity, stability requirements are being dealt with more explicitly than in the old codes, in which stability considerations have tended to be hidden in the 'safe stress' presentation. The proper application of the code depends on an improved understanding by engineers of the criteria of elastic stability. In this volume the aim has been to present a coverage of the more important aspects of structural stability, using only sufficient mathematics to allow a proper understanding, and relating the exposition to the treatment given in codes. Throughout, the aim has been towards a physical understanding of the phenomena involved, since it is only by acquiring such an understanding that an engineer can have some confidence that he will not be caught unawares by the phenomenon of buckling and catastrophic loss of load-carrying capacity. The subject of structural stability is both fascinating and of great importance, and it is hoped that this volume will demonstrate the truth of this.

1978   M. R. Horne

Preface

A study of the evolution of steel structures reveals a consistent trend towards lighter construction. This is true not only for large expensive and prestigious structures, such as long span bridges, but also for more commonplace developments such as industrial buildings. Among the reasons for this are changing visual requirements producing more elegant and cleaner lines, developments in structural form, improvements in the methods used for connecting both structural and non-structural elements, and the use of members with new cross sectional shapes, often in higher strength steels. As structures become more slender so problems associated with stability assume an increasing importance in the design and behaviour of such structures. This situation is reflected in the changing nature of the various codes of practice which are intended to provide guidance to the structural steelwork designer. Examination of these codes reveals a steady increase in the number of clauses whose function is to guard against failure by buckling of the structure as a whole, of an individual member or of an element of a member. In order to understand, and hence correctly and sensibly comply with such clauses, it has become essential for designers to possess a sound understanding of stability.

Several textbooks already exist in which the principles and the theory of instability of structures are presented. However, these are frequently regarded as being too academic for the practising engineer who is more concerned with the effect of instability problems on his designs. The authors have been conscious of this fact when preparing this text. Fundamental principles have been stated, and where necessary basic governing equations derived. Each of the topics has then been developed to demonstrate how theory may be utilised to provide design rules and calculation techniques suitable for design office use. Throughout emphasis has been placed upon hand techniques as the authors feel that a sound understanding is best achieved from worked examples many of which are included in the text. In this way it is hoped that the reader will not only obtain an understanding of fundamental principles but will also appreciate the links with code rules and design office practice.

The first chapter provides a descriptive introduction to several of the many forms of stability problems associated with steel structures. A selection of these form the subject matter for chapters 2, 3 and 4. Thus chapter 2 discusses the flexural buckling of isolated columns, thereby acting as a prelude to chapter 4 which explains how the behaviour of complete frameworks is influenced by the presence of compression members. Chapter 3 examines the way in which instability effects influence the behaviour and design of beams. Each chapter is complemented by a general bibliography to which the reader anxious for further information on a particular topic is referred. However, since the text is intended to be self-contained no system of detailed references has been included.

The preparation of any book necessarily involves the authors in discussions with many people. Whilst engaged upon the present text we have benefited from talking to colleagues in both the academic sphere as well as to engineers engaged in day-to-day design. Several of these conversations have taken place at courses and meetings organised by Constrado. In matters concerning the production of the text we have been greatly assisted by the facilities provided by the Department of Civil and Structural Engineering at the University of Sheffield. We are also grateful for the advice offered by our publishers and finally we acknowledge our gratitude to our consulting editor, Professor Horne, for his helpful guidance throughout the gestation period of this volume.

P.A.K.
D.A.N.

Notation

Chapter 1

b	Breadth of a plate element
c	Stability function—carry over factor
E	Young's modulus
I	Moment of inertia (second moment of inertia)
L	Member length
M_{AB}	Moment at end A on member AB
P	Load
P_Y	Squash load
s	Stability function—stiffness
t	Thickness
W	Load
W_{cr}	Elastic critical load
α	Scalar quantity
β	Scalar quantity
δ	Deflection
Δ	Deflection
θ_A	Rotation of A
θ_B	Rotation of B

Chapter 2

A	Cross sectional area
b	Effective eccentricity
c	Stability function—carry over factor
E	Young's modulus
e	Eccentricity
F	Force
h	Distance of extreme fibre of a section from neutral axis
I	Moment of inertia
K_b	Stiffness of bracing
K, K_2	Load factors from BS 153 and BS 449 respectively
k	Stiffness
L	Member length
ℓ	Effective length

Chapter 2 — continued

M	Moment
M_{AB}	Moment at end A on member M_{AB}
M_A	Total moment at A
m	Stability function—sway magnification factor
P	Axial load
P_E	Euler load
P_{cr}	Elastic critical load
$P_1, P_2, P_3 \ldots$	Loads defined by $\pi^2 EI/L^2$; $(2\pi)^2 EI/L^2$; $(3\pi)^2 EI/L^2 \ldots$
p_{ac}, p_c	Permissible stresses from BS 153 and BS 449 respectively
r	Radius of gyration
s	Stability function—stiffness
s''	Stability function—stiffness of member pinned at its remote end
u	Parameter defined by $\alpha L/2$
W	Load
Z	Elastic section modulus
α	Parameter defined by $\sqrt{(P/EI)}$
η	Imperfection parameter
θ	Angle of rotation
λ	Slenderness ratio ℓ/r
μ	Parameter defined by kL/P_E
σ	Normal stress
σ_{cr}	Elastic critical stress
σ_E	Euler critical stress
σ_R	Maximum compressive residual stress
σ_Y	Yield stress

Chapter 3

A	Area
B	Total breadth of flange
b	Breadth of a plate element
C	Torsional rigidity
D	Depth of section
d	Clear depth of web
E	Young's modulus
G	Shear modulus
h	Distance between flange centroids
I_f	Second moment of area of a flange
I_x	Major second moment of area
I_y	Minor second moment of area
I_w	Warping constant
J	Torsion constant
k, k_1, k_2	Effective length factors
K_b	Stiffness of bracing
L	Span
ℓ	Effective length

Chapter 3 — continued

M	Moment
$\overline{M}$	Equivalent uniform moment
M_b	Buckling resistance moment
M_{cr}	Elastic critical moment
M_f	Moment resisted by one flange
M_p	Fully plastic moment
M_Y	Moment at first yield
M_ξ, M_η, M_ζ	Components of applied moment
n	Slenderness correction factor
p_b	Bending strength
p_{bc}	Permissible bending stress
p_{cr}	Elastic critical stress
p_Y	Yield stress
Q	Applied torque
r_y	Radius of gyration about the minor axis
S_x	Plastic section modulus
s	Shape factor
T	Flange thickness
t	Thickness of plate element
u	Buckling parameter or displacement in the x-direction
V_f	Flange shear force
v	Slenderness factor or displacement in the y-direction
x	Torsional index
Z_x	Elastic section modulus
β	Ratio of end moments
γ	Factor allowing for influence of in-plane deflections
λ	Slenderness ratio
λ_{LT}	Lateral-torsional slenderness ratio
λ'_{LT}	Effective lateral-torsional slenderness ratio
η	Measurement of initial imperfection for Perry formula
η_{LT}	Imperfection used in beam application of Perry formula
ϵ_Y	Yield strain
ϕ	Twist

Chapter 4

c	Stability function—carry over factor
DC	Distribution coefficient
E	Young's modulus
F	Force
h	Horizontal distance, storey height
I	Moment of inertia
K''	Stiffness parameter
k	Member stiffness defined by EI/L
k_{BB}	Σk for all beams meeting at bottom of column
k_{BT}	Σk for all beams meeting at top of column

Chapter 4 — continued

k_{BOTTOM}	Hardy Cross distribution coefficient
k_{TOP}	Hardy Cross distribution coefficient
L	Length
ℓ	Effective length
M	Moment
M_{bal}	Balancing moment
M_R	Restoring moment
m	Stability function—sway magnification factor
n	Stability function—no shear; near end
o	Stability function—no shear; other end
P	Axial load
P_{cr}	Elastic critical load
P_E	Euler load
P_f	Failure load
P_p	Squash load
p_y	Material design strength
p_y'	Reduced material design strength
S	Spring stiffness
s	Stability function—stiffness
s''	Stability function—stiffness of a member pinned at its remote end
T	Total joint stiffness
u	Horizontal displacement
v	Vertical distance
W	Load
W_{cr}	Elastic critical load
α	Ratio of λ_c/λ_F
Δ	Deflection
λ	Load factor
λ_{cr}	Load factor at elastic critical load
λ_F	Load factor at failure
λ_P	Load factor at rigid plastic collapse
θ	Rotation
ρ	Ratio of axial load to Euler load
ρ_{cr}	Ratio of axial load to Euler load at elastic critical loading
ϕ	Sway index

The Nature of Instability

1.1 Introduction

Any structural design must satisfy two basic criteria which are:
(i) strength: the ability of the structure to bear the imposed loading and;
(ii) stiffness: the structure shall be sufficiently stiff so as not to distort more than is permissible.

To satisfy himself that his proposed structure meets the above requirements a designer relies heavily upon his analyses. Stresses and deformations at working loads are normally computed using linear elastic techniques. A rigid plastic approach may be employed to enable the ultimate load carrying capacity of the structure to be estimated. Both of these approaches neglect the influence of stability which may significantly affect both of these criteria in structures which are themselves slender or which contain members, or elements of members, which are slender and subject to compressive stresses.

Before proceeding further it is necessary to define exactly what is meant by the term stability as used in this text. It does not refer to the resistance of a structure to overturning as a rigid body (normally due to the action of lateral forces). Such behaviour is well known and may be readily analysed by simple statics. Nor is this text concerned with the problems associated with the structural integrity of the component parts of a structure which require that the designer considers carefully the way in which the forces are to be transmitted and also pays due regard to detailing. This text discusses the less well known internal instabilities which arise as a result of compressive actions.

This form of instability is a phenomenon which appears in many differing guises and is latent whenever compressive stresses are present. Its effect is to decrease the load carrying capacity of a structure (i.e. a reduction in strength) and also to increase the deformation (i.e. a reduction in stiffness). It is therefore important that designers are familiar with some of the more common modes of instability which can occur so that they may be better equipped to pursue their art. This has particular relevance in the current trends which are tending to produce more highly stressed structures of ever more slender proportions coupled with a reduction in the stiffness of infill panels and cladding, which may well have been responsible for some of the previous generation of structures performing satisfactorily.

To establish the nature of instability consider the two columns of Fig. 1.1,

Design for Structural Stability

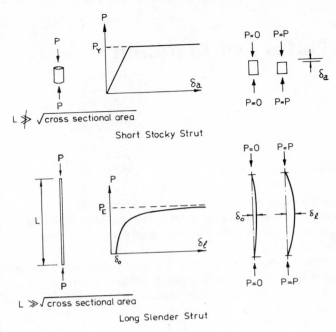

Fig. 1.1

each of identical cross-section, one short and therefore stocky, the other long and therefore slender. Under the influence of an increasing axially applied load P the stocky strut will shorten axially by some amount δ_a which will initially be proportional to the applied load. At some load, P_Y, the load divided by the cross-sectional area will attain the yield stress and the material will deform plastically. The strut 'squashes' and P_Y is aptly termed the squash load. Upon removal of the load the strut retains some permanent distortion.

By contrast the slender strut will fail in a completely different manner. If the lateral deflection at midheight is considered as a characteristic deformation, with an initial imperfection value of δ_0 under zero axial load, then, as the axial load is applied, so the lateral deflection will grow at an increasing rate until, at a certain load, the deformation will become large and the strut will fail by moving perpendicular to the line of thrust. The resulting behaviour is summarised in the lower load-deflection plot of Fig. 1.1. Provided the deformation has not been allowed to become excessive the strut will return to its original form upon removal of the load.

Clearly these two modes of failure are quite distinct. For the short stocky strut the material yields, the failure load being dependent upon the yield stress. For the slender strut the failure is elastic in nature and the phenomenon is known as Euler instability. The critical load is completely independent of the yield stress being dependent solely upon the modulus of elasticity E, the moment of inertia I, the member length L and the end conditions. This will be dealt with more fully in Chapter 2. Elastic instability may be defined as the

Fig. 1.2

condition where the structure under consideration has no tendency to return to its initial position when slightly disturbed, even when the yield stress is assumed to be infinitely large.

The same form of instability can be seen occurring in the photograph of Fig. 1.2, in which an initially flat plate is loaded along its two ends, the two remaining edges being unrestrained. The strut fails at a low load as the resistance to bending out of the plane of the plate (proportional to the minimum value of EI) is very small.

If the plate is folded at right angles along the vertical centre-line the minimum bending stiffness is radically increased and failure occurs in a different mode. Figure 1.3 shows the distortion. The two planes of the angle deflect into Euler type buckles; the amplitude of the buckle being greatest at the free edges and virtually zero at the fold. A horizontal section at midheight of the strut reveals that the cross-section itself does not distort significantly but is rotated in plan relative to the ends. This mode is essentially torsional in nature and may be considered to be initiated by lack of support at the free edges.

The mode may be inhibited by introducing additional folds to give the lipped angle strut of Fig. 1.4. Here the strut has been so proportioned that failure occurs due to local instability in which the outstands lose their stiffness before any other mode occurs. The significant parameters in this case are E, the modulus of elasticity, and b/t, the ratio of the breadth of the outstand to the material thickness. This form of instability is normally avoided in practice by limiting the maximum value of b/t so that yielding will occur prior to buckling. Figure 1.5 shows the characteristic wave pattern occurring in the flange of an I

Fig. 1.3

Fig. 1.4

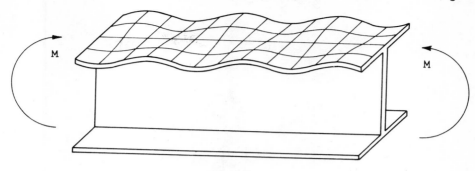

Fig. 1.5

beam and Fig. 1.6 illustrates web buckling due to shear. To fully appreciate these phenomena an understanding of plate buckling is required and in the case of webs, post-buckled stiffness. This is beyond the scope of this monograph.

The simple slender cantilever of Fig. 1.7 exhibits another form of instability—that of lateral torsional buckling in which the cantilever bends about its horizontal axis and also twists. This is dealt with in Chapter 3.

Thus far instability of a single member has been considered in isolation but structures are continuous. Consider the simple triangulated framework of Fig. 1.8 in which all of the joints are pinned. This frame will become unstable when the Euler load is attained in one member and that member is the only one which will exhibit significant distortion; in this case it is the inclined member. The horizontal tie remains straight, merely rotating as a whole to maintain continuity at the loaded joint which translates due to the shortening of the buckled inclined member. If however the members are rigidly connected at the loaded joint (Fig. 1.9) then the load necessary to cause instability is enhanced as the horizontal member which is framed into the loaded joint restrains the rotation at that end of the inclined member, and therefore affects the stiffness and also effective length of that member (see Chapter 2).

Instability is now dependent on the properties of more than one single component or one single member. In the more extensive frame of Fig. 1.10 it can be observed that all members distort (to a greater or a lesser extent) when the frame buckles.

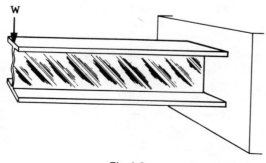

Fig. 1.6

Fig. 1.7

For a triangulated framework the distorted shape is basically composed of joint rotations but when sway frames are considered the freedom to sway may radically reduce the critical loads. A typical mode of distortion corresponding to this form of instability is indicated in the frames of Fig. 1.11 from which it can be observed that sway plays an important role even when the frame and the loading are symmetrical. This is also true for the frame of Fig. 1.12(a). The photograph of Fig. 1.12(b) shows the frame at a loading just below the elastic critical value. The corresponding diagram clarifies the distorted shape (or critical mode) from which it can be observed that the columns are tending to behave as cantilever struts over the entire height of the frame and that their slopes are being restrained by the rotational resistance of the beams at every floor level. The photograph of Fig. 1.12(c) shows the frame provided with sway bracing and carrying more than double the load without too much distress. The mode of elastic instability (which would eventually occur at about three and a half times the elastic critical load of the unbraced frame) is shown in the diagram of Fig. 1.12(c) and consists essentially of joint rotations only—a no-sway mode.

Yet another form of instability which may occur is snap-through buckling which is potentially most dangerous in shallow arches of various forms, e.g. an

Fig. 1.8(a)

Fig. 1.8(b)

Fig. 1.9(a)

Fig. 1.9(b)

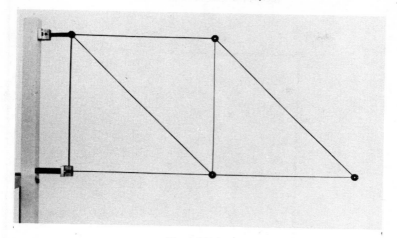

Fig. 1.10(a)

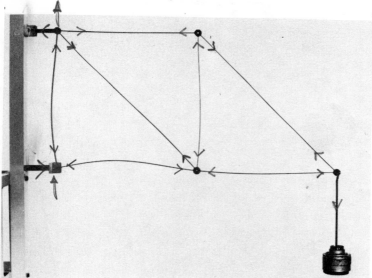

Fig. 1.10(b)

arch rib or the low-pitch tied portal of Fig. 1.13 which has been subject to a point load at the apex (omitted from the photograph for clarity). Snap-through buckling is frequently associated with large changes of geometry which often occur violently under dead loading conditions as the structure moves from one geometrical equilibrium position to another.

Having seen some of the physical manifestations of instability it is now appropriate to consider more objectively their implications on structural performance. Imagine a structure subjected to some proportional loading system defined by a load parameter W and let a characteristic distortion of the structure be δ. For the sake of clarity the frame of Fig. 1.14 will be considered in which α

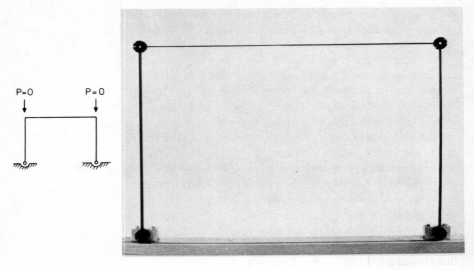

Fig. 1.11(a)

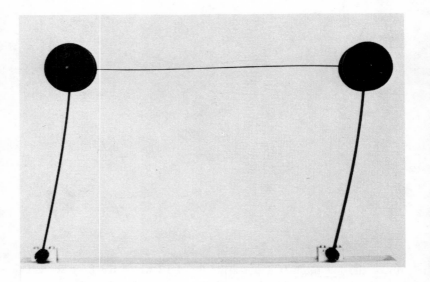

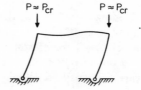

Fig. 1.11(b)

and β are simple constants and the lateral movement at beam level is taken as the characteristic distortion δ. With appropriate modifications the reasoning which follows may be applied to any structure and no restriction is implied.

The simplest form of approach which may be employed to predict structural response is linear elastic analysis, that is W is proportional to δ over an unrestricted range giving the straight line OA of Fig. 1.15. Such methods include normal moment distribution, most energy methods and the conventional slope deflection analysis. This is known as first-order linear elastic analysis. Normally a restriction is placed on the analysis—it being applicable only until the attainment of the yield stress in the structure.

If the presence of instability is recognised then W_{cr}, the elastic critical load, is an upper bound to the failure load of the frame and, for some load W, the deflections will be larger than those predicted by linear elastic theory (at a value of W equal to $W_{cr}/2$ the deflections will be approximately twice the values predicted by OA). Such methods are described in this book and include use of the familiar slope deflection equation but incorporating the non-linear coefficients s and c, i.e.

$$M_{AB} = \frac{EI}{L}\left\{ s\theta_A + sc\theta_B - \frac{s(1+c)\delta}{L} \right\} \tag{1.1}$$

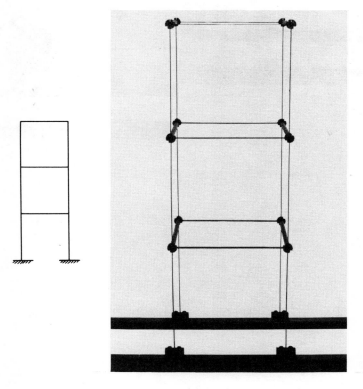

Fig. 1.12(a)

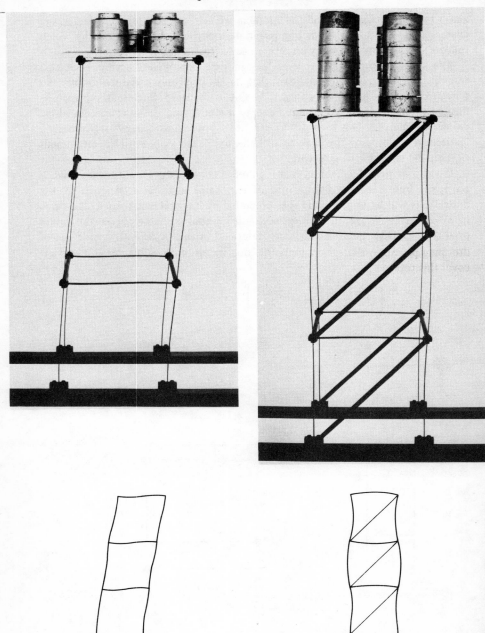

Fig. 1.12(b)   Fig. 1.12(c)

where M_{AB} is the moment at A on member AB

θ_A and θ_B are the rotations at A and B respectively

δ is the sway of B relative to A

EI and L are the flexural rigidity and length of AB, and s and c are stability functions defined by Merchant and tabulated by Livesley and Chandler.

This is known as a second-order elastic analysis and leads to the curve OBC in Fig. 1.15 which is tangential to the linear elastic characteristic at the origin and asymptotic to the line $W = W_{cr}$.

In rigid-plastic analysis it is assumed that no deformation occurs until the formation of the final hinge which converts the structure into a mechanism and the resulting load deflection plot is ODE. If the effects of the change of geometry as the mechanism distorts are included the portion DE will not be a horizontal straight line. For the frame of Fig. 1.14 compressive actions predominate and the inclusion of this effect (known as the $P - \Delta$ effect) results in the mechanism curve ODFG (which is frequently and graphically described as the drooping plastic collapse characteristic). Most texts on plastic analysis and design cover this topic.

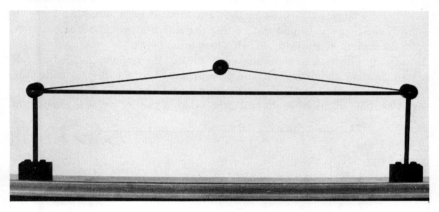

Fig. 1.13(a)

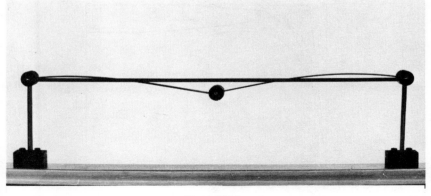

Fig. 1.13(b)

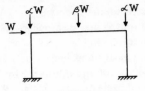

Fig. 1.14

More sophisticated analyses which are more suited to a computer-oriented approach begin with a combination of linear elastic assumptions coupled with the concept of plastic hinges. It is customary to assume unit form factor and the resulting load-deflection characteristic is then a series of straight lines, the slope changing abruptly as successive hinges form, as is indicated by OHJKL. Essentially the slope of each section represents the linear stiffness of the frame with frictionless pins inserted at locations where plastic hinges have formed at the load level considered.

If instability effects are recognised then the straight lines of the previous analysis become a series of intersecting curves, each intersection corresponding to a discrete change of stiffness at the formation of each successive hinge. If extended, each curved component of this characteristic would become asymptotic to the current elastic critical at that load level.

In order to illustrate these idealisations consider the response of the frame of Fig. 1.14. Initially the deflection would be predicted by the characteristic OBC but at a certain value of load (corresponding to M on Fig. 1.15) the bending moment at some location would reach the value of the fully plastic moment of

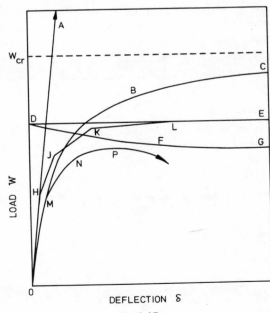

Fig. 1.15

resistance at that section. The structure now has a new linear elastic stiffness (defined by the slope of HJ on the linear elastic plastic characteristic) and hence a new non-linear stiffness which includes the effects of axial loads. Both of these stiffnesses are determined considering the structure modified by the insertion of a frictionless pin at the location of the plastic hinge as it is considered that the section is unable to resist any additional moment for an increment of load. The structure also has a new reduced elastic critical load, frequently referred to as the deteriorated elastic critical load, W'_{cr}, being the critical load of the deteriorated structure, i.e. with pins inserted at the locations of the plastic hinges. Consequent upon the reduction in the elastic critical load is the fact that the characteristic would, for increasing loads and in the absence of the formation of further hinges, become asymptotic to the line $W = W'_{cr}$. However at some value of load, corresponding to point N of Fig. 1.15, a further hinge will form and a new regime will then operate.

Once again the structure is modified with an additional freely rotating hinge being inserted. The elastic stiffness and the elastic critical load are again reduced and the process is repeated until the load parameter becomes equal to the elastic critical load of the structure in the current deteriorated condition. This may arise at the formation of the hinge as the critical load reduces from a value above the load corresponding to the hinge formation to a value which is below. It may also arise as the loading proceeds if the current critical load lies between the values of load which causes two successive hinges to form.

The true behaviour of the structure will still not be obtained due to the large number of mathematical idealisations still present even in this quite sophisticated approach. These assumptions include such items as initial stresses, initial geometrical imperfections, material variability, finite member and joint sizes, real joint stiffnesses (which may vary with increasing load) and so on.

It falls upon the engineer to assess the degree of refinement required in his calculations: it is clearly pointless to perform an analysis which gives results 'accurate' to say, 1% if the input parameters (loadings, material properties, etc.) are only known to 10%. Similarly it is not appropriate to use an analysis which incorporates a form of action which will not be significant in the structure under consideration.

Instability is probably one of the least understood areas in structural behaviour and design. It is hoped that this publication will enable the reader to identify, and to improve his appreciation of, this problem and hence to execute his designs with greater awareness.

CHAPTER TWO

In-Plane Instability of Columns

2.1 Introduction

The overall instability of a single member subject to an axial compressive force can occur in three distinct fashions:

(i) Euler buckling about the minor principal axis;
(ii) Euler buckling about the major principal axis;
(iii) Pure torsional instability.

It is commonly assumed when dealing with such a problem that instability will always occur by Euler buckling about the minor principal axis, but this is erroneous. Indeed it has been shown by Kappus (as early as 1938) that the elastic critical load of a thin-walled column of general open shape is a function of all of the three modes listed above, i.e. as low as, or lower than, any (i) (ii) or (iii) individually. For sections which have two axes of symmetry, all three modes are independent and, for most rolled or fabricated sections which have relatively thick walls, the elastic critical load is the Euler load corresponding to minor axis flexure. Torsional buckling however is still likely with thin-walled cruciform sections and cold-formed sections are liable to torsional buckling.

2.2 Overall Euler Buckling

Historically the foundations of instability theory were laid down by Euler in 1744 when he showed that the cantilever column of Fig. 2.1 could not carry a load in excess of $\pi^2 EI/4L^2$. It is more usual however to consider first the case of a 'perfect' pin-ended column.

Fig. 2.1

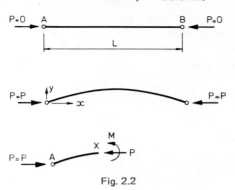

Fig. 2.2

2.3 The Perfect Pin-Ended Column—The Classical Approach

Consider a perfectly straight, pin-ended column which is subject to a perfectly axial load and which is constrained to distort only in one plane—the xy plane. Consideration of the equilibrium of the portion AX of such a member which is moved to a slightly distorted position as indicated in Fig. 2.2 gives the equation

$$M + Py = 0 \qquad (2.1)$$

Using the Euler—Bernoulli bending equation

$$M = EI \frac{d^2 y}{dx^2} \qquad (2.2)$$

(which is valid for small deflections only) gives

$$\frac{d^2 y}{dx^2} + \frac{Py}{EI} = 0 \qquad (2.3)$$

This is a second-order differential equation the solution of which is of the form

$$y = C_1 \sin \sqrt{\frac{P}{EI}} x + C_2 \cos \sqrt{\frac{P}{EI}} x \qquad (2.4)$$

where C_1 and C_2 are arbitrary constants. Putting $\alpha^2 = P/EI$ and differentiating twice with respect to x gives

$$\frac{d^2 y}{dx^2} = -C_1 \alpha^2 \sin \alpha x - C_2 \alpha^2 \cos \alpha x \qquad (2.5)$$

It can be seen quite readily that this solution satisfies the original Eq. 2.3. It now remains to determine the two arbitrary constants.

The boundary conditions are

(i) At A: when x = 0, y = 0 giving $C_2 = 0$;
(ii) At B: when x = L, y = 0 giving $C_1 \sin \alpha L = 0$.

Condition (ii) may be satisfied in two ways.

Either $C_1 = 0$ and the strut does not deflect

or $\sin \alpha L = 0$

This only occurs when $\alpha L = n\pi$

$$\text{i.e. when} \quad P = \frac{n^2 \pi^2 EI}{L^2} \tag{2.6}$$

where $n = 1,2,3 \ldots$

Hence

(i) For loads less than $P = \pi^2 EI/L^2$ the deflection y is always zero;

(ii) For the load $P = \pi^2 EI/L^2$ the deflection y is undefined—i.e. it is uncontrolled.

This means that at this load (which is called the Euler load) the application of any disturbing influence such as an end moment or a lateral force will not be resisted by the member, no matter how small this disturbing force, at least within the bounds of small deflection theory. In this imperfect world this is the limit of the practical range as far as entire members of all but the most unusual structural applications are concerned.

The resulting load deflection characteristic for this perfect idealised case is shown as the dashed line of Fig. 2.3.

There exists an infinite number of additional solutions to the differential equation satisfying the boundary conditions. They are given by

$$P_n = \frac{n^2 \pi^2 EI}{L^2} \qquad \text{where n is an integer.}$$

The only one of practical significance is that obtained from $n = 1$. The others (given by $n = 2,3,4 \ldots$) correspond to the higher modes as indicated in Fig. 2.4 and only occur if additional constraints are inserted, e.g. bracing to prevent lateral deformation of the midlength of the pin-ended strut making the case corresponding to the $n = 2$ solution.

Had an expression for curvature more accurate than Eq. 2.2 been employed the solution of the differential equation would have resulted in the chain-dotted characteristic shown on Fig. 2.3. This is a line which commences from $P = P_E$ at $\delta = 0$ but which rises as deformation increases. This demonstrates that struts do

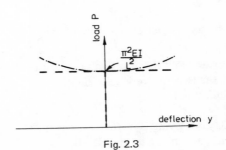

Fig. 2.3

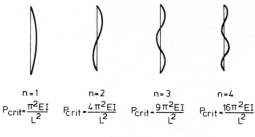

Fig. 2.4

have some post-buckling strength but for all practical purposes this enhanced load-carrying capacity is negligible due to the flatness of the curve in the range of acceptable deformations.

2.4 Initial Bow in the Pin-Ended Column

It is impossible to attain the idealised condition described above and real columns usually contain two types of geometric imperfection, namely:

(i) the member itself is not perfectly straight;
(ii) the load is not applied axially.

The latter effect causes additional bending in the member and will be dealt with under the heading of 'beam columns' in Section 2.5. The initial shape of the structural column will obviously vary from member to member and it is convenient to express this unloaded shape in the form of a series expansion

$$y_0 = a_1 \sin(\pi x/L) + a_2 \sin(2\pi x/L) + \cdots \cdots \tag{2.7}$$

where y_0 is the perpendicular distance from the chord joining the two ends of the member measured to the centre line of the member as indicated in Fig. 2.5.

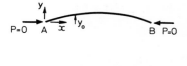

Fig. 2.5

This form of series expansion can be used to describe any distorted shape. It can be shown that the deflection of the centroidal axis of the strut, y, after the application of a load P is given by

$$y_c = \frac{a_1}{1 - P/P_1} \sin \frac{\pi x}{L} + \frac{a_3}{1 - P/P_2} \sin \frac{2\pi x}{L} + \cdots \cdots$$

$$+ \frac{a_n}{1 - P/P_n} \sin \frac{n\pi}{L} + \cdots \cdots \qquad (2.8)$$

The derivation of this expression is shown below.

Consider the equilibrium of the portion A of the strut of Fig. 2.5, and taking moments about X

$$M + Py = 0$$

Now

$$M = EI \left\{ \frac{d^2 y}{dx^2} - \frac{d^2 y_0}{dx^2} \right\} \text{ for } \left(\frac{dy}{dx} \right)^2 \ll 1 \text{ and } \left(\frac{dy_0}{dx} \right)^2 \ll 1$$

Then

$$\frac{d^2 y}{dx^2} + \frac{Py}{EI} = \frac{d^2 y_0}{dx^2}$$

Putting $\alpha^2 = P/EI$ it can be shown by back substitution that the solution is

$$y = C_1 \sin \alpha x + C_2 \cos \alpha x + \frac{a_1}{1 - P/P_1} \sin \frac{\pi x}{L} + \frac{a_2}{1 - P/P_2} \sin \frac{2\pi x}{L} + \cdots \cdots$$

where $P_1 = \pi^2 EI/L^2$; $P_2 = (2\pi)^2 EI/L^2$, etc.
It now remains to evaluate the two arbitrary constants C_1 and C_2 using the two boundary conditions

(i) at $x = 0$; $y = 0$ giving $C_2 = 0$;
(ii) at $x = L$; $y = 0$ giving $C_1 \sin \alpha L = 0$.

Thus either $C_1 = 0$ or $\sin \alpha L = 0$
When $\sin \alpha L = 0$, $P = n^2 \pi^2 EI/L^2$, i.e. this solution is only valid at the Euler load and its multiples.
At loads less than P_E

$$y = \frac{a_1}{1 - P/P_1} \sin \frac{\pi x}{L} + \frac{a_2}{1 - P/P_2} \sin \frac{2\pi x}{L} + \frac{a_3}{1 - P/P_3} \sin \frac{3\pi x}{L} + \cdots \cdots$$

Of the terms in $a_1, a_2 \ldots$ etc. all of those with even subscripts vanish when $\sin n\pi = 0$ whilst the remaining terms have values of plus and minus unity alternately. Thus the expression for the central deflection becomes

$$y_c = \frac{a_1}{1 - P/P_1} - \frac{a_3}{1 - P/P_3} + \frac{a_5}{1 - P/P_5} - \cdots \cdots \qquad (2.9)$$

Now clearly as P tends to P_1 the first term of the expression tends to infinity as its denominator approaches zero. The denominators of the subsequent terms

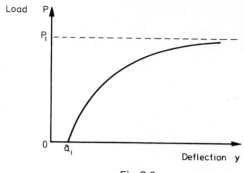

Fig. 2.6

however do not approach zero. For example since $P_3 = 9P_1$, when P reaches P_1 the denominator of the third term is $1 - P_1/9P_1 = 0.89$.

The physical significance of each of the terms may be explained as follows. The first term describes the effect on the central deflection of the strut bending into a single half sine wave; the second term accounts for the influence of the strut distorting into one and a half sine waves, the third into two and a half sine waves and so on. Note that the terms in a_2, a_4 are absent as these correspond to modes which have nodes at the midlength of the column.

When plotted into Fig. 2.6 this expression for the deflection, Eq. 2.9, describes a curve which is asymptotic to the Euler load P_1. As P approaches P_1 (P greater than say $P_1/2$), the first term dominates and at loads less than $P_1/2$ the error in neglecting all but the first term is likely to be quite small.

The factor $1/(1 - P/P_1)$ is known as the amplification factor and, as will be seen later, is frequently a very useful parameter for estimating the effects of instability on stresses and deflections.

2.5 Column Subjected to Lateral Loads—The Beam Column

In an elastic member subjected to bending action it is well known that in the absence of axial load, the stresses and deformations are directly proportional to the magnitude of the imposed lateral loads. Computations may be conducted using the undistorted shape of the member when setting up the equations without incurring any significant errors. If however, an axial load is also present, as in Fig. 2.7, this approach is no longer valid.

Due to the action of the lateral loading a lateral deformation, Δ, will result. Due to this lateral deformation an addition bending moment $P\Delta$ will occur, causing additional deformation which will in turn result in a further increase in

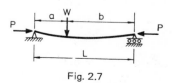

Fig. 2.7

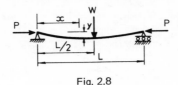

Fig. 2.8

moment. Solution of the problem by this iterative intuitive approach is thus not practical and it is necessary to reformulate and solve the fundamental governing equations. Even for a simple situation such as that indicated in Fig. 2.7 this results in a quite lengthy computation. If however, the lateral load W is at midspan then, because of symmetry, the algebra involved is of more manageable proportions and this example may conveniently be used to illustrate the method and to demonstrate the correlation between the resulting 'exact' solution and a much simpler approximation presented by Timoshenko.

Consider the doubly pin-ended beam column of Fig. 2.8 which carries a central disposed lateral point load W. Let the length of the member be denoted by L and its flexural rigidity (assumed constant throughout) by EI. In the absence of the axial thrust P the central deflection y_{max} would be given by $WL^3/48EI$.

Returning to the problem where the axial load P is present it is necessary to write down an expression for the bending moment at some general location distant x from the left-hand end of the member.

For $0 \leqslant x \leqslant L/2$

$$M_x = \frac{Wx}{2} + Py.$$

Substituting $M = -EI \dfrac{d^2 y}{dx^2}$ and putting $\alpha^2 = P/EI$ as before

$$\frac{d^2 y}{dx^2} + \alpha^2 y + \frac{W\alpha^2}{2P} = 0 \tag{2.10}$$

The solution of this equation can be shown (by back substitution) to be

$$y = C_1 \sin \alpha x + C_2 \cos \alpha x - \frac{Wx}{2P} \tag{2.11}$$

The constants of integration (A and B) may now be found by considering the following boundary conditions:

(i) At x = 0; y = 0 giving $C_2 = 0$;
(ii) At x = L/2; by symmetry dy/dx = 0.

Note that x = L; y = 0 does not constitute a boundary condition as the initial expression is only valid within the left-hand half of the beam.
Differentiating Eq. 2.11

$$\frac{dy}{dx} = C_1 \alpha \cos \alpha x - \frac{W}{2P} \tag{2.12}$$

Evaluating at $x = L/2$ and equating to zero gives

$$A = W / \left(2P\alpha \cos \frac{\alpha L}{2}\right).$$

Hence the deflected shape is given by

$$y = \left[W / \left(2P\alpha \cos \frac{\alpha L}{2}\right)\right] \sin \alpha x - Wx/2P \tag{2.13}$$

Thus the central deflection is

$$y_{max} = [W/(2P\alpha)] \tan \frac{\alpha L}{2} - \frac{\alpha L}{2} \tag{2.14}$$

At first sight this expression appears to bear little similarity to $WL^3/48EI$ but the correlation may be rapidly established by considering the substitution

$$u = \frac{\alpha L}{2} \quad (\text{or } \alpha = 2u/L)$$

$$u^2 = \frac{\alpha^2 L^2}{4} = \frac{L^2}{4} \frac{P}{EI} \quad \text{i.e. } P = 4u^2 EI/L^2$$

Hence

$$y_{max} = \frac{WL^3}{48EI} \left[\frac{3(\tan u - u)}{u^3} \right] \tag{2.15}$$

It can readily be appreciated that the term in the square brackets represents the amplification (due to the presence of the axial load P) of the linear elastic deflection calculated ignoring the effect of the axial load.

The problem can be seen to be very similar to the solution of the simple column problem. The differences are the additional term in the differential equation plus the modified boundary conditions – here requiring symmetry for the simple solution. Were the load not centrally disposed then two equations of the form of Eq. 2.10 would have been necessary, one being valid for the beam to the left of the load and the other for the beam to the right of the load. The four boundary conditions necessary to evaluate the four arbitrary constants resulting from the pair of differential equations would have been derived from the two end conditions plus deflection and slope continuity at the location of the point load. Alternatively McCauly type brackets may be used (with suitable care being taken to preserve their correct meaning during subsequent integrations).

Using either of the above methods the deflected shape for the more general case depicted in Fig. 2.7 can be shown to be

$$y = \frac{W \sin \alpha b \sin \alpha x_1}{\alpha P \sin \alpha L} - \frac{Wbx_1}{PL} \tag{2.16}$$

This expression is valid for $0 \leqslant x_1 \leqslant a$ with x_1 measured inwards from the left-hand end of the beam. A similar expression defines the shape of the

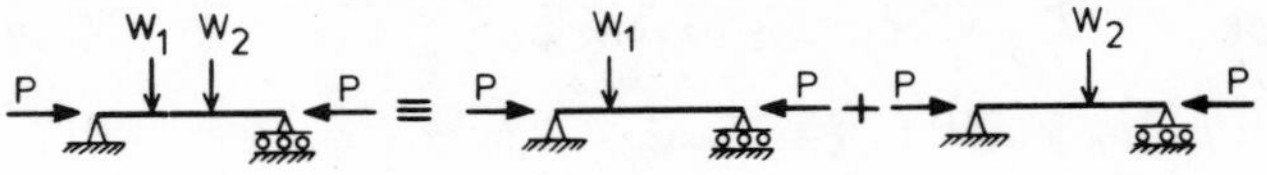

Fig. 2.9

remainder of the beam

$$y = \frac{W \sin \alpha a \sin \alpha x_2}{\alpha P \sin \alpha L} - \frac{W a x_2}{PL} \tag{2.17}$$

where x_2 is measured inwards from the right-hand end of the beam.

It should here be noted that the principle of superposition may (with suitable care) be applied to find deflections (and bending moments) due to a number of point loads as these quantities are linear functions with respect to the lateral applied loads and thus the total lateral deformation due to a series of point loads can be found as the sum of the deflections resulting from each individual point load acting together with the full axial load as indicated in Fig. 2.9.

Practising engineers are accustomed to computing linear elastic deflections and are understandably reluctant to solve differential equations when an alternative approach is available. An approximate method was proposed by Timoshenko which is based on using the amplification factor developed earlier in Section 2.4 when the effects of geometric imperfections on the deflection of a pin-ended column were examined. Essentially the approximation states that the deflection y of a point on a beam-column under an axial thrust P may be estimated by calculating the linear elastic deflection y_e at that point ignoring P, and then multiplying y_e by the amplification factor $1/(1 - P/P_{cr})$ where P_{cr} is the value of P at which the column would become elastically unstable. Considering again the example of Fig. 2.8, the quantity u was defined earlier and can be shown to be a simple function of P/P_E as follows:

$$u = \frac{L}{2} \sqrt{\left(\frac{P}{EI}\right)} = \frac{\pi}{2} \sqrt{\left(\frac{P}{P_E}\right)} \tag{2.18}$$

Figure 2.10 compares graphically the amplification factor $3(\tan u - u)/u^3$ with the simpler approximation $1/(1 - P/P_E)$ for increasing values of axial load P. Note that for this pin-ended member P_{cr} is P_E. It can readily be seen that there is very close correspondence between the two solutions. Such close agreement does not always result in high values of P/P_{cr} but it has been shown that for values of P/P_{cr} not exceeding 0·6 the error incurred in using the approximation will not exceed 2%.

In the problem of Fig. 2.8 the bending moment at any point is given by

$$M = \frac{Wx}{2} + \frac{W \sin \alpha x}{2\alpha \cos \dfrac{\alpha L}{2}} - \frac{Wx}{2} \tag{2.19}$$

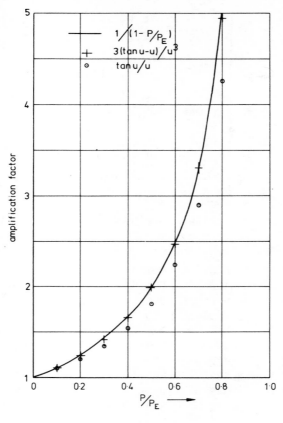

Fig. 2.10

which at midspan becomes

$$M_{max} = \frac{W \tan \frac{\alpha L}{2}}{2\alpha} = \frac{WL}{4} \left[\frac{\tan u}{u} \right] \qquad (2.20)$$

Once again this contains two terms, the first being the bending moment which would occur in the absence of P and the term in brackets which represents an amplification factor. Comparison of this factor with increasing axial loads is also shown in Fig. 2.10 together with the correlation with the simple amplification factor $1/(1 - P/P_E)$. Tabulated values appear in Fig. 2.12.

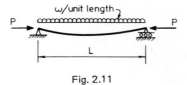

Fig. 2.11

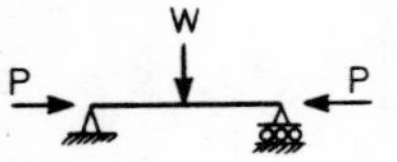

$$y_{max} = \frac{WL^3}{48EI}\left[\frac{3(\tan u - u)}{u^3}\right] = \frac{WL^3}{48EI}\phi_1$$

$$M_{max} = \frac{WL}{4}\left[\frac{\tan u}{u}\right] = \frac{WL}{4}\phi_2$$

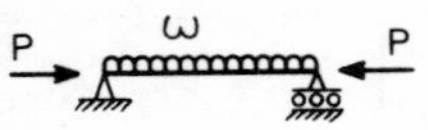

$$y_{max} = \frac{5wL^4}{384EI}\left[\frac{12}{5}\frac{(2\sec u - 2 - u^2)}{u^4}\right] = \frac{5wL^4}{384EI}\phi_3$$

$$M_{max} = \frac{wL^2}{8}\left[\frac{2(\sec u - 1)}{u^2}\right] = \frac{wL^2}{8}\phi_4$$

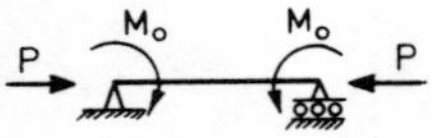

$$y_{max} = \frac{M_0 L^2}{8EI}\left[\frac{2(\sec u - 1)}{u^2}\right] = \frac{M_0 L^2}{8EI}\phi_4$$

$$M_{max} = M_0[\sec u] = M_0 \phi_5$$

P/P_E	$1/(1 - P/P_E)$	ϕ_1	ϕ_2	ϕ_3	ϕ_4	ϕ_5
0	1.000	1.000	1.000	1.000	1.000	1.000
0.1	1.111	1.109	1.091	1.107	1.114	1.137
0.2	1.250	1.247	1.205	1.253	1.258	1.310
0.3	1.429	1.420	1.350	1.430	1.441	1.533
0.4	1.667	1.657	1.545	1.663	1.684	1.831
0.5	2.000	1.982	1.815	2.000	2.029	2.252
0.6	2.500	2.477	2.223	2.502	2.544	2.884
0.7	3.333	3.303	2.901	3.347	3.407	3.942
0.8	5.000	4.943	4.253	5.013	5.124	6.057
0.9	10.000	9.876	8.308	10.040	10.290	12.420

Fig. 2.12

Estimation of the maximum bending moment by multiplying WL/4 by the simple amplification factor does not lead to an accurate assessment of the true bending moment but the approximation may still give an acceptable degree of accuracy for many purposes and at least serves as a useful check on more refined analysis.

Other problems may be treated in a similar manner, for example the beam column of Fig. 2.11 where the lateral loading is in the form of a uniformly distributed loading of w per unit length over the entire member length. It can be shown that the central deflection is given by

$$y_{max} = \frac{5wL^4}{384EI}\left[\frac{12}{5}\frac{(2\sec u - 2 - u^2)}{u^4}\right] \tag{2.21}$$

and

$$M_{max} = \frac{wL^2}{8}\left[\frac{2(\sec u - 1)}{u^2}\right] \tag{2.22}$$

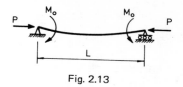

Fig. 2.13

Figure 2.12 compares the exact amplification factors within the square brackets with the simple approximation and again the close correspondence is evident.

Similarly for a column loaded by equal moments M_0, as shown in Fig. 2.13, the maximum deflection and moment can be shown to be given by

$$y_{max} = \frac{M_0 L^2}{8EI}\left[\frac{2(\sec u - 1)}{u^2}\right]$$

(2.23)

and

$$M_{max} = M_0 [\sec u]$$

(2.24)

Note that the amplification factor for y_{max} is identical with that corresponding to M_{max} for the previous load case.

Figure 2.12 shows the comparison of the amplification factors derived via solution of the differential equations with the simple amplification factor $1/(1 - P/P_E)$.

Summarising, it can be seen that for the computation of deformations the simple amplification factor can be employed with complete confidence up to quite high axial loads (say 60% of P_{cr}) whilst even for the evaluation of bending moments its usage gives a good indication of behaviour.

Complicated loading arrangements may be considered using superposition of individual loading components provided each individual effect is computed under the full axial load.

2.6 Eccentrically Loaded Column

The commonly encountered situation of a column carrying a load parallel to its longitudinal centroidal axis but at an eccentricity e is a particular case of a column subjected to end moments with both moments being equal to Pe, see Fig. 2.14. Using Eq. 2.24 the maximum moment occurring in the column may be written

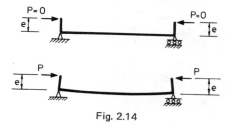

Fig. 2.14

as

$$M_{max} = Pe \sec \frac{\alpha L}{2} \tag{2.25}$$

If allowance is also made for lack of straightness as in Section 2.4 then it can readily be seen from Eq. 2.9 that an additional moment will occur giving a total maximum moment of

$$M_{max} = Pe \sec \frac{\alpha L}{2} + \frac{Pa_1}{1 - P/P_1} - \frac{Pa_3}{1 - P/P_3} + \cdots \tag{2.26}$$

This represents the situation shown in Fig. 2.15. The same result may be obtained from first principles leading to the result for the maximum deflection.

$$y_{max} = e \sec \frac{\alpha L}{2} + \frac{a_1}{1 - P/P_1} - \frac{a_3}{1 - P/P_3} + \cdots$$

Using the good approximate expansion

$$\sec \frac{\alpha L}{2} = \left\{ 1 + 0.26 \frac{P}{P_1} \right\} \bigg/ \left\{ 1 - \frac{\alpha^2 L^2}{\pi^2} \right\}$$

the expression for the central deflection becomes

$$y + e = \frac{e\{1 + 0.26P/P_1\} + a_1}{1 - P/P_1} - \frac{a_3}{1 - P/P_3} + \cdots \tag{2.27}$$

Neglecting all subsequent terms it can be seen that the first term is still quite complex and recognising that it is impractical to attempt to measure a_1 and e for all columns it is not unreasonable to ascribe an equivalent effective geometric imperfection constant b to account for all geometric imperfections.

The expression for $(y + e)$ at midheight is now $b/(1 - P/P_1)$ and it can be seen that this is the initial effective imperfection b multiplied by a simple factor $1/(1 - P/P_1)$. This is the familiar amplification factor (not to be confused with the magnification factor m of stability functions, see later) and can be used to assess deflection including the effects of instability when only the linear elastic deflections are known.

2.7 Stresses in Geometrically Imperfect Columns

As yet no reference has been made to the stress in the column, nor to the yield stress of the material. For very slender columns the limiting factor on the

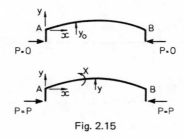

Fig. 2.15

load-carrying capacity is solely a function of the stiffness of that column derived from purely elastic analysis. It is customary in design codes to make reference to maximum stress but this is solely for convenience and sometimes leads to misconceptions. The Euler critical stress, which is entirely hypothetical, is obtained by dividing the critical load by the cross-sectional area of the column, e.g. for a pin-ended column

$$P_E = \pi^2 EI/\ell^2 = \pi^2 EAr^2/\ell^2$$

or

$$\sigma_E = \pi^2 Er^2/\ell^2 = \pi^2 E/(\ell/r)^2 \tag{2.28}$$

where σ_E is the mean stress over the cross-section at the critical load. It can be seen that this is equal to two constants divided by the square of the ratio ℓ/r which is known as the slenderness ratio and frequently denoted by the symbol λ. Figure 2.16 shows a plot of critical stress versus slenderness ratio. The region to the upper right-hand side of the curve indicates that failure will have occurred by elastic instability.

So far it has been assumed that yielding of the material does not occur. It is obvious that as ℓ/r becomes very small, i.e. stocky columns, failure will occur wholly by yielding at the squash load and the line defined by $\sigma = \sigma_Y$ is another upper limitation on the load-carrying capacity of the member. Any condition represented by a stress in excess of σ_Y is clearly inadmissible since prior failure will have occurred due to yielding at the squash load which is the product of the cross-sectional area and the yield stress.

For columns which are neither very stocky nor extremely slender (which covers the majority of practical cases) failure will occur due to a combination of the two modes and at a load which is lower than both the squash load and the elastic critical load.

In an initially imperfect column the compressive stress due to the application of the load P is made up of two components:

(i) that due to the axial load in the column, P/A, which is constant throughout the member length;

and

(ii) that due to the bending action, $P(y + e)/Z$, where $(y + e)$ is the perpendicular

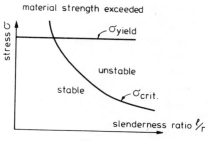

Fig. 2.16

distance between the line of action of the load P and the centroidal axis of the column and Z is the elastic section modulus of the section considered.

Again assuming that the deflection is a maximum at the midheight of the column the greatest compressive stress in the section is on the concave side and is given by

$$\sigma_{max} = \frac{P}{A} + \frac{P(y + e)}{Z} \text{ evaluated at midheight}$$

or

$$\sigma_{max} = \frac{P}{A} + \frac{P}{Z} \frac{b}{(1 - P/P_E)} \tag{2.29}$$

Putting σ_{max} equal to the yield stress σ_Y and writing $Z = Ar^2/h$ where r is the appropriate radius of gyration and h is the distance of the extreme fibre from the neutral axis

$$\sigma + \frac{\sigma h}{r^2} \frac{b}{(1 - \sigma/\sigma_E)} = \sigma_Y \tag{2.30}$$

where σ is P/A, the nominal applied stress over the cross-section, and σ_E is P_E/A. Rearranging

$$\sigma(1 - \sigma/\sigma_E) + \sigma \frac{bh}{r^2} = \sigma_Y(1 - \sigma/\sigma_E)$$

$$\sigma^2 - \sigma_E\left(1 + \frac{bh}{r^2}\right) + \sigma_Y\sigma + \sigma_E\sigma_Y = 0$$

which is a quadratic in σ the solution of which is

$$\sigma = \frac{\sigma_E(1 + \eta) + \sigma_Y}{2} - \sqrt{\left[\frac{\sigma_Y + (1 + \eta)\sigma_E^2}{2} - \sigma_Y\sigma_E\right]} \tag{2.31}$$

where η (= bh/r^2) is a non-dimensional parameter which describes the geometric imperfection.

2.8 Residual Stresses

Geometric imperfections do not account for all of the differences between the real column in a structure and the mathematical model upon which the calculations developed so far have been based. Perhaps the least recognised of these differences until the last few years has been the assumption that in an unloaded condition the material is free from both stress and strain.

The coefficient of thermal expansion of mild steel is about $1 \cdot 3 \times 10^{-5}$ per $^\circ$C and hence the strain induced due to raising the temperature of steel by 100°C is $1 \cdot 3 \times 10^{-3}$. Taking the modulus of elasticity as say 200×10^3 N/mm^2 and a yield stress of 200 N/mm^2 the strain at first yield is $1 \cdot 0 \times 10^{-3}$. Thus it can be

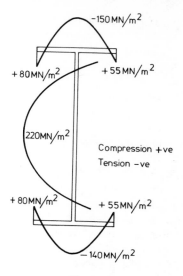

Fig. 2.17

seen that the mechanical strains which correspond to yielding at sensibly constant temperatures are greatly exceeded by unrestricted thermal expansions produced during the rolling process used in the manufacture of structural sections. Subsequent cooling does not occur uniformly through the section and it is not surprising that even in an unloaded condition structural members contain significant self-equilibrating patterns of stresses which are described as residual stresses. Similar effects occur during all thermic processes such as welding and flame cutting. Considerable research has been undertaken to measure values of residual stresses by sectioning members longitudinally and measuring the released strains. A typical 'average' pattern of residual stresses for a universal beam section is shown in Fig. 2.17.

Although the behaviour of the column is not affected whilst the material remains elastic, since the principle of superposition applies, the presence of the residual stresses will normally reduce the load at which yielding first occurs. Consider the case of a short stocky column where instability does not occur. Yielding will first occur at an axial load given by $A(\sigma_Y - \sigma_R)$ where σ_R is the maximum value of residual compressive stress. Uncontrolled squashing will however still occur at an axial load of $A\sigma_Y$. Neglecting the presence of residual stresses one would expect an abrupt transition from a wholly elastic condition to a fully plastic state whereas the change is a more gradual process. In the partially plastic condition the effective flexural rigidity of the section may be estimated by assuming that the E value for the yielded sections is zero and thus the effective flexural rigidity EI_{eff} may be written as

$$EI_{eff} = EI_{elastic\ zones\ only}$$

Thus the critical load of a pin-ended column may be written as

$$P_{cr} = \frac{\pi^2 (EI)}{L^2} \text{ elastic zones.} \tag{2.32}$$

This is described at the tangent modulus buckling load.

2.9 Real Isolated Columns and Design Codes

In recent years a number of computer-based analytical procedures, which deal more or less adequately with many of the features present in real columns, have been developed. Such methods enable research studies to be performed giving an insight into the behaviour of actual columns but their complexity clearly makes them unsuitable for direct use, i.e. routine design. Most design codes adopt the more practical approach of basing column design on a formula derived on the assumption that only one form of imperfection is present (e.g. lack of straightness, residual stresses, etc.).

In the current British Standards strut design is based on Eq. 2.31. As indicated in Section 2.7 this equation is not based on a collapse condition but on the attainment of first yield. However Robertson found that by selecting a suitable value for η a curve could be obtained which formed an approximate lower bound to the test data available at that time. The value of $\eta = 0.003\ell/r$ proposed by Robertson is still used in BS 153:1972. As a result of later work by Godfrey in 1962 BS 449:1969 uses a different value of $\eta = 0.3(\ell/100r)^2$, which gives slightly higher nominal applied stresses at low values of ℓ/r (ℓ/r less than 100).

In both cases the tables of permissible stresses actually quoted in the codes (denoted by p_{ac} in Table 4 of BS 153 and p_c in Table 17 of BS 449) are obtained by dividing the mean stress (σ of Eq. 2.31) by a load factor (denoted by K and K_2 in BS 153 and BS 449 respectively). A value of approximately 1·7 is used for both K and K_2.

A greatly increased amount of data on column behaviour has become available in recent years and this information has indicated that various types of section will perform differently when used as columns. These variations arise mainly as a result of the severity of the residual stresses present in the sections, which in turn affects the manner in which each section loses stiffness due to the spread of yield (see Section 2.8). This implies that if design is based on a single design curve then sections which are less severely affected by residual stresses will be penalised if sections which are more severely affected are not to be underdesigned. Detailed theoretical studies of this problem, in which failure loads were determined for a variety of column sections each containing representative residual stress patterns and an initial bow of $\ell/1000$, have suggested the use of three curves with a group of sections being allocated to each curve. Extensive full-scale testing programs have confirmed the theoretical work. In addition the tests demonstrated that the full squash load is attainable for short columns. Figure 2.18 gives the proposed European column curves in the non-

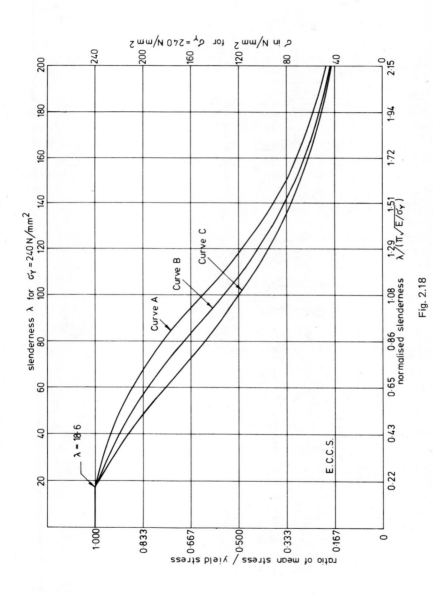

Fig. 2.18

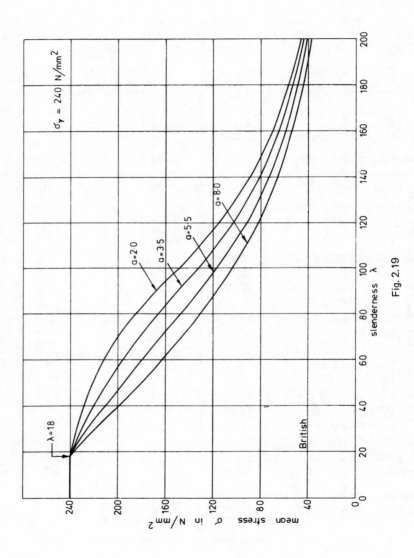

Fig. 2.19

dimensional form first proposed by the European Convention for Constructional Steelwork in which the ratio of the mean stress at failure σ to the yield stress σ_Y is plotted against a normalised slenderness $\lambda/\pi \sqrt{(E/\sigma_Y)}$. The latter parameter has the value of unity when $\sigma_{cr} = \sigma_Y$. (Remember that $\lambda = \ell/r$.)

It seems likely that column design in the new British Code for structural steelwork will be based on an adaptation of the European column curves. The approach uses a modified Perry formula to provide a very close approximation to the European column curves with the mean stress, σ, being obtained from Eq. 2.31 but with η being defined as

$$\eta = a\,(\lambda - 0 \cdot 2\pi\sqrt{(E/\sigma_Y)}) \times 10^{-3} \text{ but not less than zero.}$$

Figure 2.19 shows how by selecting suitable values for 'a' the modified Perry formula may be made to provide a close fit to the original European column curves. Also shown in Fig. 2.19 is an additional lower curve corresponding to a value of $8 \cdot 0$ which is considered to be appropriate for very heavy sections containing plate elements over 40 mm thick since such sections perform relatively poorly when used as columns. As far as the designer is concerned the only real difference involved in using the new approach as compared with BS 449:1969 is the need to select the column curve appropriate to the particular section under consideration using a selection table provided.

2.10 The End Conditions of a Column

Thus far no reference has been made to the real end conditions of the column and it has been assumed in all preceding work that the ends are pinned permitting free rotation but rigidly located with respect to sway. Rarely are these criteria satisfied in practice; normally columns will frame into other members which afford finite restraints. In such circumstances the problem of determining the elastic critical load is really a frame problem and rigidly jointed frameworks will be discussed as such in Chapter 4. There are however several idealised end conditions which can be examined using the methods and mathematics employed earlier in this chapter.

For example Euler's pioneering work started with the analysis of a column encastré at its base and free at the other end as shown in Fig. 2.20. It can be seen that the analysis developed from Fig. 2.2 is valid if Eq. 2.5 is solved using the end conditions

(i) At $x = 0$; $y = 0$ (again giving $C_2 = 0$)
(ii) At $x = 0$; $dy/dx = 0$

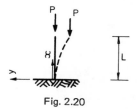

Fig. 2.20

Differentiating Eq. 2.4, and remembering that $C_2 = 0$ gives

$$\frac{dy}{dx} = C_1 \frac{P}{EI} \cos \sqrt{\left(\frac{P}{EI}\right)} x = 0 \qquad (2.33)$$

So either $C_1 = 0$ which gives an unreal solution as it implies that the column never deflects or

$$\cos \sqrt{\left(\frac{P}{EI}\right)} L = 0 \qquad (2.34)$$

which occurs when

$$\sqrt{\left(\frac{P}{EI}\right)} L = (n - \frac{1}{2})\pi$$

where n is an integer

i.e. $P = (n - \frac{1}{2})^2 \pi^2 EI/L^2$

$$= \pi^2 EI/4L^2 ; 9\pi^2 EI/4L^2 \ldots \text{etc.} \qquad (2.35)$$

The important load here is the first which is the critical load.

It can be seen that this is of the same form as the solution for the compression member with pinned ends and, had the lengths of latter case been doubled and inserted into the formula for the pin-ended strut, the same result would have been obtained as shown below. Let ℓ be the length of the pin-ended strut and L be the length of the cantilever strut.

Put

$$\ell = 2L$$

then

$$P_{cr} = \frac{\pi^2 EI}{\ell^2} = \frac{\pi^2 EI}{4L^2} \qquad (2.36)$$

as before.

The length ℓ is known as the 'effective length' of the real strut and is the length of an 'equivalent' pin-ended strut which has the same critical load. It is also the distance between adjacent points of contraflexure on the real strut or

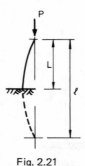

Fig. 2.21

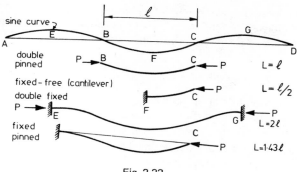

Fig. 2.22

(as in this case) on the real strut and the continuation of its deflected shape as indicated in Fig. 2.21.

The concept of effective length is extremely convenient for design as the permissible mean axial stress in any column may be determined from a single design rule (for instance Eq. 2.31 in the current BS 449) which relates to a pin-ended member. Thus in the design of an axially loaded member it is necessary to assess the effective length ℓ, and dividing by the appropriate radius of gyration, obtain the slenderness ratio ℓ/r (often denoted by the symbol λ). With this slenderness, the appropriate value of the Robertson constant a and the material design strength the mean permissible stress may be obtained. For idealised end constraints the effective length may be quickly and easily determined.

Remembering that the distorted shape of a column varies along its length in a sinusoidal manner (Eq. 2.4) and referring to the sine curve of Fig. 2.22 it can be observed that where the curve crosses the median line (A, B, C and D) there is zero curvature — zero curvature on an initially straight structural element implies zero moment which is the criterion of a pin connection. Thus the length BC represents the form of distortion for a doubly pin-ended strut. The tangents to the curve at E, F and G are parallel to the median line and thus it can be seen that FC represents the form of the distorted shape of a cantilever strut. Note that here the material of the strut within the length FC will be subject to actions which are identical to those experienced by the corresponding material in the pin-ended strut BC and both will suffer instability at the same load. However the cantilever strut FC is only half the length of BC which will become unstable at an axial load given by

$$P_{cr} = \frac{\pi^2 EI}{\ell^2}$$

The cantilever strut will become unstable at the same load $\pi^2 EI/\ell^2$ but its real length L is only $\ell/2$ and thus its critical load is given by

$$P_{cr} = \frac{\pi^2 EI}{4L^2}$$

It can also be seen that EBFCG represents the distorted form of a strut encastré at both ends where the ends are constrained from sway. Here the effective length is now one half of the real length and thus

$$P_{cr} = \frac{4\pi^2 EI}{L^2}$$

A strut which is built-in at one end but pinned at the other and where sway is prevented is slightly more difficult to distinguish. Identification of C as the location of the pin end is simple enough, but E does not mark the encastré support as some sway would then be involved. This end is located by drawing the tangent to the curve from point C which osculates the sine curve slightly to the right of E. This distorted shape can now be seen to satisfy the end constraints which are (i) that one end is pinned and (ii) that the tangent at the other (encastré) end passes through the pinned end. The effective length can be found as $1 \cdot 43L$ by carefully plotting the sine curve in the region of E and graphically constructing the tangent. Alternatively it could be found by inserting the relevant end conditions into Eq. 2.4 which leads to an awkward transcendental equation $\tan kL = kL$. There is a further approach using stability functions which is introduced later. This technique enables this particular problem to be solved with only two lines of simple calculation and is sufficiently versatile to deal with many other problems.

The concept of assessing effective lengths is widely employed in design codes to deal with a variety of end constraints. For example Clause 31(a) of BS 449 gives the effective lengths to be adopted for five combinations of idealised end restraints for use in structures with simple connections. These are listed below together with the theoretical equivalent values assuming fully rigid connections. (It is likely that one or two further categories will be introduced into the new code.)

END CONDITIONS	BS 449	THEORETICAL
1. Effectively held in position and restrained in direction at both ends	$\ell = 0 \cdot 7L$	$\ell = 0 \cdot 5L$
2. Effectively held in position at both ends and restrained in direction at one end	$\ell = 0 \cdot 85L$	$\ell = 0 \cdot 7L$
3. Effectively held in position at both ends but not restrained in direction	$\ell = 1 \cdot 0L$	$\ell = 1 \cdot 0L$
4. Effectively held in position and restrained in direction at one end, and at the other partially restrained in direction but not held in position.	$\ell = 1 \cdot 5L$	$\ell = ?$

5. Effectively held in position $\ell = 2 \cdot 0L$ $\ell = 2 \cdot 0L$
   and restrained in direction
   at one end but not held in
   position or direction at the
   other end.

In the first two cases the recommended effective lengths are greater than the theoretical ones, implying practical load-carrying capacities below the theoretical loads. This reflects the difficulty of obtaining ends which are perfectly encastré. The third case corresponds exactly with the theoretical result; presumably on the grounds that any small degree of freedom to sway is offset by some rotational restraint. The final restraint condition (number 5) corresponds exactly to the theoretical value on the grounds that if one end is completely free, full fixity must be provided at the other. The remaining condition (number 4) referring to some intermediate partial restraint in direction is difficult to justify on a theoretical basis but as will be seen later quite a moderate rotational restraint is sufficient to cause a significant increase in the elastic critical load of a compression member which has one pinned end. However, although the effective length is reduced from $2 \cdot 0L$ in condition number 5 to $1 \cdot 5L$ (i.e. by 25%) it must be remembered that this corresponds to an increase in the elastic critical load of $(2 \cdot 0)^2 : (1 \cdot 5)^2$ or $4 : 2 \cdot 25$, i.e. about 78%.

It can therefore be appreciated that whilst great precision with respect to permissible stresses for pin-ended members appears in current design codes the adoption of an effective length to convert a real column into the equivalent pin-ended member is imprecise.

The major problem is that the end restraints which can occur for a real column are infinitely variable from full fixity to completely free. Furthermore the restraint for example against in-plane rotation is dependent not only upon the flexural rigidity and length of any restraining member but also upon the axial load in this member and the degree of fixity present at its remote end including flexibility of the connections. Thus even column instability problems are actually frame problems. Attempts to overcome this difficulty are sometimes made by estimating the locations of the points of contraflexure. Although it is possible to find points of contraflexure in an analysis using linear elastic methods these will not in general be the locations which will exist at the elastic critical load, even assuming fully rigid joints, and thus the result will be erroneous.

It is somewhat surprising that, in view of the extent of knowledge about the behaviour of an isolated pin-ended column and the refinements adopted in the rules for its design, the assessment of effective lengths is left to what essentially amounts to at best an estimate and at worst a guess. For example, following the guidance of Appendix D of BS 449, one might estimate the effective length of a compression member in a frame as being in the range $0 \cdot 7L$ to $0 \cdot 85L$. This does not however represent a region of uncertainty of about 20%. Consider the implication on the load-carrying capacity if stability effects dominate. Translating the two effective lengths into elastic critical loads gives

(i) $\ell_1 = 0.7L$ $P_{cr_1} = \dfrac{\pi^2 EI}{\ell_1^2} \bigg/ \dfrac{\pi^2 EI}{0\cdot49L^2}$

(ii) $\ell_2 = 0\cdot85L$ $P_{cr_2} = \dfrac{\pi^2 EI}{\ell_2^2} = \dfrac{\pi^2 EI}{0\cdot723L^2}$

or

$$\frac{\ell_1}{\ell_2} = 1\cdot21$$

but

$$\frac{P_{cr_1}}{P_{cr_2}} = 1\cdot48$$

Thus whilst the region of uncertainty in effective lengths is about 20% translation into load-carrying capacity in the case of slender struts can increase this uncertainty to a staggering 50%.

Fortunately there are improved methods of assessing effective lengths for fully rigid jointed frames using various stability analyses and one using stability functions is developed in the remainder of this chapter.

2.11 Stability Functions

2.11.1 Introduction

The elastic critical loads of isolated columns may be derived by setting up and solving differential equations as in the previous section. There is however an alternative approach which involves the use of stability functions which are, in effect, solutions of the differential equations in a general tabulated form.

The merits of these functions are most pronounced when used to find elastic critical loads of frameworks but they do enable simple problems to be solved rapidly and derivation of known accepted results leads to increased confidence when they are employed in more complicated situations.

The standard slope-deflection equation relating end moments and end distortions of a beam is well known:

$$M_{AB} = \frac{EI}{L}\left\{4\theta_A + 2\theta_B - \frac{6\delta}{L}\right\} \tag{2.37}$$

where

  M_{AB} is the moment at A on member AB
  EI is the flexural rigidity of member AB
  L is the length of member AB

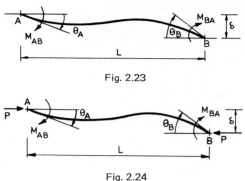

Fig. 2.23

Fig. 2.24

θ_A is the rotation of A
θ_B is the rotation of B

and

δ is the sway of member AB

These quantities are illustrated in Fig. 2.23. Moments and rotations are taken as positive when clockwise and δ is positive if it tends to rotate the member as a whole in a clockwise sense. The equation is a form of linear elastic analysis and does not include the effects of axial thrust in the member.

Perhaps somewhat less well known is the same equation which includes the effect of axial loads as shown in Fig. 2.24:

$$M_{AB} = \frac{EI}{L} \left\{ s\theta_A + sc\theta_B - \frac{s(1+c)\delta}{L} \right\} \tag{2.38}$$

Comparison with the normal slope-deflection equation reveals that the constant coefficients 4, 2 and 6 have been replaced by s, sc and s(1 + c) respectively. The values of s, the member stiffness, and c, the carryover factor, vary solely with the ratio of axial load in the member to the Euler load of that member. The ratio is commonly denoted by the symbol ρ and can readily be calculated from

$$\rho = PL^2/\pi^2 EI$$

2.11.2 Derivation of Stability Functions

Consider the strut of Fig. 2.25 which is

(i) encastré at one end and pinned at the other;
(ii) subject to an axial load P;
(iii) subject to a moment M at its pinned end.

Let the length of the member be L and its flexural rigidity be EI. Define the moments M_{AB} and M_{BA}, the rotation θ_A and the axes x,y as indicated in the diagram. Let s and c be defined by $M_{AB} = sEI\theta_A/L$ and $M_{BA} = cM_{AB}$.

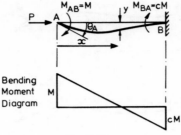

Fig. 2.25

At some point distant x from A the bending moment M_x is given by

$$M_x = M\left\{1 - \frac{(1 + c)x}{L}\right\} + Py = -EI\frac{d^2y}{dx^2} \tag{2.39}$$

or

$$\frac{d^2y}{dx^2} + \frac{Py}{EI} = \frac{M}{EI}\left\{\frac{x(1 + c) - L}{L}\right\} \tag{2.40}$$

Now

$$P_E = \frac{\pi^2 EI}{L^2}$$

or

$$EI = P_E L^2/\pi^2$$

$$\therefore \frac{d^2y}{dx^2} + \frac{P}{P_E}\frac{\pi^2}{L^2}y = \frac{M\pi^2}{P_E L^2}\left\{\frac{x(1 + c) - L}{L}\right\} \tag{2.41}$$

Put

$$\bar{K} = \frac{\pi}{L}\sqrt{\left(\frac{P}{P_E}\right)}$$

Then

$$(D^2 + \bar{K}^2)y = \frac{\bar{K}^2 M}{P}\left\{\frac{x(1 + c) - L}{L}\right\} \tag{2.42}$$

The solution of this differential equation is

$$y = \frac{M}{PL}\left\{x(1 + c) - L\right\} + A\sin\bar{K}x + B\cos\bar{K}x \tag{2.43}$$

Putting

$$\alpha = \frac{\bar{K}L}{2}$$

$$y = \frac{M}{PL}\left\{x(1 + c) - L\right\} + A\sin\frac{2\alpha x}{L} + B\cos\frac{2\alpha x}{L} \tag{2.44}$$

Consider next the boundary conditions

(i) $x = 0$; $y = 0$ $\therefore B = M/P$

(ii) $x = L$; $y = 0$ $\therefore = -\dfrac{M}{P}\left\{\dfrac{c + \cos 2\alpha}{\sin 2\alpha}\right\}$

(iii) $x = L$; $\dfrac{dy}{dx} = 0$ gives, after some working,

$$c = \frac{\sin 2\alpha - 2\alpha}{2\alpha \cos 2\alpha - \sin 2\alpha} \qquad\qquad (2.45)$$

(iv) $x = 0$; $\dfrac{dy}{dx} = \theta_A$ and using $M_{AB} = sEI\theta_A/L$

gives, after considerable manipulation,

$$s = \frac{\alpha(1 - 2\alpha \cot 2\alpha)}{\tan \alpha - \alpha} \qquad\qquad (2.46)$$

Now

$$\alpha = \frac{\bar{K}L}{2} = \frac{\pi}{L}\frac{L}{2}\sqrt{\left(\frac{P}{P_E}\right)} = \frac{\pi}{2}\sqrt{\rho}$$

i.e. dependent solely upon the ratio of axial load to Euler load, and hence s and c may be tabulated for all struts in terms of ρ.

The preceding analysis relates only to rotations without sway.
Sway may be considered as follows:
Consider a strut AB

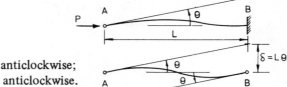

(i) Rotate end A through θ anticlockwise;
(ii) Rotate end B through θ anticlockwise.

Assuming that θ is small this is equivalent to a small sway deformation of magnitude $L\theta$ with no rotation at A or B. Positive sway is

OPERATION	MOMENT AT A	MOMENT AT B	
Initial condition	0	0	
(i) Rotate A thro θ	$-s\theta EI/L$	$-sc\theta EI/L$	−ve as θ is anticlockwise
(ii) Rotate B thro θ	$-sc\theta EI/L$	$-s\theta EI/L$	
(i) + (ii) or sway deformation δ	$-s(1 + c)\theta EI/L$	$-s(1 + c)\theta EI/L$	

defined as that tending to rotate the member as a whole in a clockwise sense and therefore the moment due to sway is $-s(1 + c)\theta EI/L$.

Noting that $\delta = L\theta$ the sway moment is $-s(1 + c)EI\delta/L^2$ and the modified slope-deflection equation is:

$$M_{AB} = \frac{EI}{L} \left\{ s\theta_A + sc\theta_B - \frac{s(1+c)\delta}{L} \right\} \tag{2.47}$$

Full tabulated values of s, c and $s(1+c)$ appear in *Tables of Stability Functions* by Livesley and Chandler, which also includes a series expansion which is particularly useful in computer programs using these stability functions. Selected tabulated values of these functions are listed in the Appendix and the diagram of Fig. 2.26 shows how s and c vary with the axial load parameter ρ. The derivation of these functions seems involved but fortunately their use is quite simple and a few examples are given to demonstrate this simplicity.

Consider first the situation where the ends of a uniform member are constrained against moving perpendicular to the line joining the ends of the member — a no-sway case.

2.12 Prismatic Members without Sway

Example 2.1

Find the elastic critical load of a slender 152 × 152 × 23 kg/m U.C. 6·00 metres high. It is effectively encastré at its base and pinned in position at its upper end. It has no additional restraint in the plane parallel to the flanges and a load, P, is applied to the head of the column as indicated in Fig. 2.27.

As there is no restraint in the weaker principal plane (bending about the yy axis) deformation will occur in this plane. The appropriate Euler load, P_E, may therefore be calculated as follows:

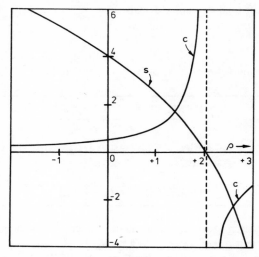

Fig. 2.26

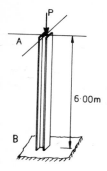

Light bracing preventing sway
but providing negligible rotational restraint

A

6·00m

B

Fig. 2.27

From tables $I_{yy} = 403$ cm^4 and, taking E as 207 kN/mm^2,

$$P_E = \frac{\pi^2 \times 207 \times 403 \times 10^4}{6\cdot00^2 \times 10^6} = 229 \text{ kN}$$

Recalling that the definition of the elastic critical load is the load at which the strut cannot resist any applied disturbance, it is clear that to test for instability any convenient disturbance may be selected. In this case an obvious disturbance is one in which end A is rotated through some angle θ_A. When this distortion is applied a moment M_A of $EIs\theta_A/L$ occurs at A and a moment M_B of $EIsc\theta_A/L$ is carried over to B. End B is encastré and therefore these moments are those which virtually arise due to the distortion. If the strut is stable there will be positive resistance to the distortion,

i.e. $M_A = EIs\theta_A/L > 0$

If the strut is unstable this quantity will be negative. The elastic critical load is attained when the resistance becomes zero

i.e. $M_A = EIs\theta_A/L = 0$

Clearly if EI/L is zero there is no strut and this solution is trivial, and if θ_A is zero there is no disturbance. Thus the condition for instability is that $s = 0$. Referring to the curve of s in Fig. 2.26 it can be seen that this first occurs when $\rho = 2\cdot05$, i.e. when $P = 2\cdot05P_E = 2\cdot05 \times 229$ kN = 469 kN. Thus the setting up and solving of a differential equation is replaced by a simple computation.

Example 2.11

Find the elastic critical load of an identical strut with pin supports at both ends.

Once again it is convenient to test for stability by rotating end A through θ_A. If this is done whilst holding B fixed then the moments are the same as in the previous problem. Now, however, it is impossible for B to resist the moment $EIsc\theta_A/L$ and this end will rotate by an amount θ_B such that the net moment at the pin end is zero,

i.e. $M_B = EIsc\theta_A/L + EIs\theta_B/L = 0$

or $\theta_B = -c\theta_A$

When B rotates through this angle an additional moment of $EIsc\theta_B/L$ arises at A giving a total moment at A of

$$M_A = (EI/L)(s - sc^2)\,\theta_A$$

For instability the term $s(1 - c^2)$ must vanish and as this quantity appears frequently in stability problems it is tabulated and often denoted by s''. It is shown plotted on Fig. 2.28 which makes clear that the lowest critical load occurs when $\rho = 1\cdot00$,

i.e. $P_{cr} = P_E$ or $P_{cr} = 229$ kN.

It can be seen that for a single column values of EI and L need not be included in the calculation of the critical value of ρ. Also the values of θ_A and θ_B are purely relative and again may be omitted. Tabulated versions of Example 2.11 are shown below which include and exclude the constants EI, L and θ.

WITH CONSTANTS

	M_A	M_B
Rot A	$\dfrac{EI}{L}s\theta_A$	$\dfrac{EI}{L}sc\theta_A$
Rot B	$\dfrac{EI}{L}sc\theta_B$	$\dfrac{EL}{L}s\theta_B$
Balance B	Put $\theta_B = -c\theta_A$	

$$M_A = \frac{EIs\theta_A}{L} - \frac{EIsc^2\theta_A}{L}$$

$$= \frac{EIs}{L}(1 - c^2)\theta_A$$

For instability $s(1 - c^2) = 0$

i.e. $\rho = 1\cdot00$

WITHOUT CONSTANTS

(i) Rot A   (ii) Rot B

s	sc
sc	s

Balance B by (i) $-$ c(ii)

$$M_A \propto s - sc^2 = s(1 - c^2)$$

For instability $s(1 - c^2) = 0$

i.e. $\rho = 1\cdot00$

2.13 Prismatic Members with Sway

In the absence of axial loads (the situation of Fig. 2.29) the shear equilibrium equation is well known as

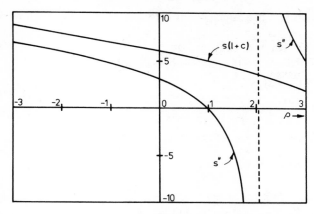

Fig. 2.28

$$F = -\frac{M_{AB} + M_{BA}}{L} \tag{2.48}$$

It is necessary to re-examine the shear equation in the presence of axial thrust. Taking moments about one end of the axially-loaded member of Fig. 2.30 leads to the relationship:

$$F = -\left\{\frac{M_{AB} + M_{BA}}{L}\right\} + \frac{P\delta}{L} \tag{2.49}$$

where F is the shear force at A or B necessary to maintain equilibrium in the presence of end moments M_{AB} and M_{BA} in a member of length L subject to an end thrust P and with a sway δ. Comparison of Eqs. 2.48 and 2.49 shows that the $P\delta/L$ term is the only modification and, if either $P = 0$ or $\delta = 0$, then the two equations are identical. There is a simple way of dealing with this additional term but it requires that the distortion of the member under consideration be

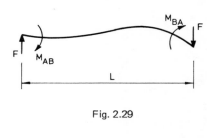

Fig. 2.29

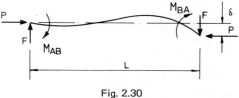

Fig. 2.30

considered in its two basic component deformations as set out below:

(i) End rotations without sway; $\delta = 0$
Equation 2.49 reduces to the normal shear equation as $\delta = 0$.

$$F = -\frac{M_{AB} + M_{BA}}{L} \qquad\qquad (2.48) \text{ (repeated)}$$

(ii) Pure sway without rotations; $\theta_A = \theta_B = 0, \delta \neq 0$.
Equation 2.49 can be shown to reduce to

$$F = -\frac{M_{AB} + M_{BA}}{mL} \qquad\qquad (2.50)$$

where m (which is known as the magnification factor) is a function solely of the ratio of P and P_E for the member and which can therefore be tabulated in the same manner as s and c. The way in which it varies is shown in Fig. 2.31. Note that when $\rho = 0$ it has the value of unity and that as ρ increases m increases demonstrating that the member end moments necessary to resist a lateral force F rise. Indeed as ρ approaches $1 \cdot 00$, m tends to infinity showing that the end moments required to resist a small lateral force become very large.

The Shear Equation under Pure Sway Conditions

Consider a strut subject to pure sway deformations as shown in Fig. 2.30. Now

$$\theta_A = \theta_B = 0, \text{ so } M_{AB} = M_{BA} = M = \frac{EI}{L}\left\{\frac{-s(1+c)\delta}{L}\right\}$$

Thus

$$\delta = -ML^2/EIs(1+c)$$

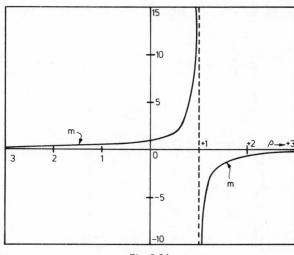

Fig. 2.31

Considering equilibrium of AB and taking moments about A:

$$M_{AB} + M_{BA} + FL + P\delta = 0$$

Thus

$$FL = -2M + \frac{PML^2}{EIs(1+c)} = -2M + \frac{PML^2}{EIs(1+c)}\frac{\pi^2EI}{P_EL^2}$$

$$= -2M\left\{1 - \frac{\pi^2\rho}{2s(1+c)}\right\}$$

or

$$F = -\frac{2M}{mL}$$

where $m = 1\Big/\left\{1 - \frac{\pi^2\rho}{2s(1+c)}\right\}$

It must be remembered that m represents a magnification factor for the propping force necessary for equilibrium for *pure sway*. For *zero sway* however the normal shear equation applies (m = 1) and for this reason care must be exercised in keeping sway and rotational distortions separate until the necessary shear equilibrium equations have been established. Thereafter of course any combination may be formed.

Example 2.III

Find the critical load for the strut of Example 2.1 where the ends are clamped against rotation but where one end is free to move laterally, as indicated in Fig. 2.32.

Here $\theta_A = \theta_B = 0$ and for a lateral deflection δ of B

$$M_{AB} = M_{BA} = -\frac{EIs(1+c)\delta}{L^2}$$

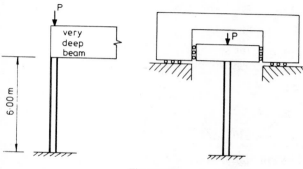

Fig. 2.32

The shearing force necessary to cause this deformation is

$$F = -\frac{M_{AB} + M_{BA}}{mL} = \frac{2EIs(1 + c)\delta}{mL^3}$$

Taking this lateral deflection as the test distortion, for instability the shearing force necessary to produce this is zero and therefore either

$$s(1 + c) = 0 \text{ or } m = \infty$$

Inspection of Figs. 2.28 and 2.31 shows that the latter occurs at the lowest load when $\rho = 1 \cdot 00$. Thus $P_{cr} = P_E = 229$ kN. Below is the diagrammatic representation of the computation.

$$F = \frac{2s(1 + c)}{m}$$

$M = -s(1 + c)$

$M = -s(1 + c)$

For instability F = 0 or

$$\frac{s(1 + c)}{mL} = 0 \tag{2.51}$$

i.e. $\rho_{cr} = 1 \cdot 00$

Example 2.IV

Find the critical load of the same column when fixed at one end and free at the other — a cantilever strut as shown in Fig. 2.33. The solution to this problem can be easily deduced from effective lengths but the example serves to further illustrate the use of stability functions which have wider ranges of application.

This case enters the range where combinations of distortion patterns are necessary and the process is as follows:

(i) Consider sway without rotation:

$$M_{AB} = M_{BA} = -s(1 + c) \, EI\delta/L^2$$

$$F = 2s(1 + c)EI\delta/mL^3$$

Note that m appears in the shear equation (pure sway).

(ii) Consider rotation of A without sway:

$$M_{AB} = sEI\theta_A/L$$

$$M_{BA} = scEI\theta_A/L$$

$$F = -s(1 + c)EI\theta_A/L^2$$

Note that m does not appear in the shear equation (pure rotation).

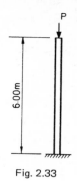

Fig. 2.33

Two suitable test distortions are possible, either a small rotation of A or a small lateral movement of A. Here the former is selected and the stiffness of the strut to the distortion is measured as M_A. As end B is fixed it is not necessary to monitor the moments arising here. The values of M_{BA} are required only to determine the shear force, F, for each component distortion. As there is no means of providing a force F at A, F must be zero and therefore

$$2s(1 + c)EI\delta/mL^3 - s(1 + c)EI\theta_A/L^2 = 0$$

or

$$\delta = \frac{mL}{2}\theta_A$$

Therefore the net moment at A due to a rotation of θ_A when the column sways until no propping force is present is given by

$$M_{AB} = -s(1 + c)EI\left(\frac{mL}{2}\right)\theta_A/L^2 + sEI\theta_A/L$$

$$= \left\{s - \frac{ms(1 + c)}{2}\right\}\frac{EI\theta_A}{L}$$

For instability

$$s - ms(1 + c)/2 = 0 \tag{2.52}$$

This must be solved by trial and error:

ρ	0	0·1	0·2	0·25	0·30
$s - ms(1 + c)/2$	1·0	0·647	0·235	0·001	−0·260

Thus $P_{cr} = 0·25P_E = 57$ kN.

Diagrammatically the computation may be represented thus:

(i) Sway of A relative to B without rotation:

$$F = \frac{2s(1 + c)}{mL}$$

F ⟶ A $M = -s(1 + c)$

$M = -s(1 + c)$
B

(ii) Rotation of A without sway:

$$F = \frac{-s(1 + c)}{L}$$

F ⟶ A $M = s$

$M = sc$
B

Combine (i) and (ii) to eliminate F, i.e. (i) × m/2 − (ii)

$$F = 0 \quad M_{AB} = -\frac{ms(1 + c)}{2} + s$$

for instability $s - ms(1 + c)/2 = 0$ as before.

Alternatively using the effective length concept, $\ell = 2\cdot 0L$, i.e. $12\cdot 00$ m. Thus

$$P_{cr} = 57 \text{ kN}$$

The final case with idealised end constraints is now examined.

Example 2. V

Find the elastic critical load of the same column with both ends effectively fixed against rotation and where sway is prevented as indicated in Fig. 2.34.

In this problem no end distortions are possible. However, the solution may be obtained by considering an additional point C at the midheight of the column. Intuitively from symmetry it can be seen that the distortion pattern which will occur will be one in which C moves laterally without rotation. The pattern of

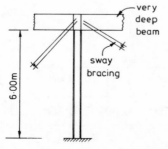
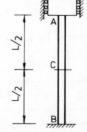

Fig. 2.34

moments associated with this distortion is identical with that of Example 2.III, but with the distortion pattern of the earlier example occurring over a length of 3·00 metres. The critical value of ρ was found to be 1·00 and this must be converted to a critical load. It must be remembered that the value of ρ is that of member AC or CB and not that of the original column AB.

$$\rho_{AC\ cr} = 1\cdot00 = P_{AC}/P_{EAC} = P_{AB}\Bigg/\left\{\frac{\pi^2 EI_{AB}}{(L_{AB}/2)^2}\right\}$$

or

$$1.00 = P_{AB}/4P_{EAB}$$

Therefore instability occurs when $P_{AB} = 4P_{EAB} = 4 \times 229 = 916$ kN.

The examples so far have considered only prismatic members with idealised end constraints. Normally in real situations columns frame into other members with the result that the actual problem is a frame problem. This will be dealt with later. The basic problem associated with non-prismatic sections can, however, be dealt with in relation to the behaviour of the isolated column.

2.14 Non-prismatic Members

Example 2.VI

Consider the specific example of a stanchion ABC encastré at its base, carrying a vertical roof load of P at A and a vertical crane load of 5P at B as indicated in Fig. 2.35. The relative lengths of AB and BC are 1:2·5 and their inertias are in the ratio of 1:9·38, giving relative stiffnesses I/L of 1:3·75. Suppose that the stanchion roof joint is not free to translate laterally. This is an assumption which would be justified only if bracing were incorporated to prevent this.

Let the suffixes 1 and 2 denote that the particular parameter is that associated with the length AB or BC respectively.

It is necessary to establish values of the parameters listed below:

Member	Axial Load	Inertia	Length	Euler Load	Relative ρ	Relative k
AB	P	I_1	L_1	$\dfrac{\pi^2 EI}{L_1^2} = P_{E1}$	1	1
BC	6P	$9\cdot38I_1$	$2\cdot5L_1$	$\dfrac{\pi^2 E9\cdot38I_1}{(2\cdot5L_1)^2} = 1\cdot5P_{E1}$	4	3·75

Consider in turn the possible distortions which are θ_A, θ_B and δ_B.

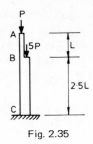

Fig. 2.35

(i) Rotation of A through θ_A with B clamped (see Example 2.II):

$$M_{AB} = \frac{EI}{L_1} s_1 \theta_A = Ek_1 s_1 \theta_A$$

$$M_{BA} = \frac{EI}{L_1} s_1 c_1 \theta_A = Ek_1 s_1 c_1 \theta_A$$

$$H_B = \frac{EI}{L_1} \frac{s_1(1 + c_1)\theta_A}{L_1} = Ek_1 s_1 (1 + c_1) \theta_A / L_1$$

This is a pattern of moments which will be factored subsequently and combined with other patterns. As indicated before there is therefore no real need to continually include all the constants, e.g. E and θ_A, and this pattern may thus be simplified to

$$M_{AB} = k_1 s_1; M_{BA} = k_1 s_1 c_1 \text{ and } H_B = k_1 s_1 (1 + c_1)/L_1.$$

(ii) Rotation of B through θ_B with A clamped gives (see Example 2.II):

$$M_{AB} = k_1 s_1 c_1$$
$$M_{BA} = k_1 s_1$$
$$M_{BC} = k_2 s_2$$
$$M_{CB} = k_2 s_2 c_2$$
$$H_A = -k_1 s_1 (1 + c_1)/L_1$$
$$H_B = k_1 s_1 (1 + c_1)/L_1 - k_2 s_2 (1 + c_2)/L_2$$
$$H_C = k_2 s_2 (1 + c_2)/L_2$$

At this point it is worthwhile noting that it is no longer necessary to monitor the moments on each member meeting at a joint, as only the total moment at each joint is required. Also the moments and propping forces at the fixed end are not subsequently required, nor is the propping force at A which is carried by the bracing. The necessary information then reduces to

$$M_A = k_1 s_1 c_1$$

$$M_B = k_1 s_1 + k_2 s_2$$

$$H_B = k_1 s_1 (1 + c_1)/L_1 - k_2 s_2 (1 + c_2)/L_2$$

(iii) Sway of B through δ_B with A and B fixed against rotation gives

$$M_A = k_1 s_1 (1 + c_1)/L_1$$

$$M_B = k_1 s_1 (1 + c_1)/L_1 - k_2 s_2 (1 + c_2)/L_2$$

$$H_B = 2k_1 s_1 (1 + c_1)/(m_1 L_1^2) + 2k_2 s_2 (1 + c_2)/(m_2 L_2^2)$$

Note that the magnification factor m appears in the H_B expression as pure sway is involved, whereas in (i) and (ii) above it was omitted as those distortions were purely rotational.

The problem now reduces to finding a pattern of moments corresponding to one distortion, say a small lateral movement of C and evaluating the lateral force associated with this distortion with no moment restraint at A or B. The patterns are manipulated, successively eliminating the moments at A and B, and it is then found that the lateral force H_B necessary to cause a small lateral deflection of B with A and B free to rotate until there is no net out of balance moment at either point is given by:

$$H_B \propto \frac{2k_1 s_1 (1 + c_1)}{mL_1^2} - \frac{k_1 s_1 (1 + c_1)^2}{L_1^2} + \frac{2k_2 s_2 (1 + c_2)}{m_2 L_2^2}$$

$$- \left\{ \frac{k_1 s_1 (1 - c_1^2)}{L_1} - \frac{k_2 s_2 (1 + c_2)}{L_2} \right\}^2 \bigg/ \left\{ k_1 s_1 (1 - c_1^2) + k_2 s_2 \right\}$$

Note that the magnitude of the lateral movement is not required as for instability the force necessary to cause any small amount of lateral movement is zero. Putting this expression equal to zero and multiplying by L_1^2, dividing by k_1 and remembering that $L_1/L_2 = 0·4$ and $k_1/k_2 = 3·75$ the following expression is obtained:

$$H_B \propto 2s_1 (1 + c_1)/m_1 - s_1 (1 + c_1)^2 + 1·2s_2 (1 + c_2)/m_2$$

$$- \frac{\{s_1 (1 - c_1^2) - 1·5s_2 (1 + c_2)\}^2}{s_1 (1 - c_1^2) + 3·75s_2} = 0 \qquad (2.53)$$

This is a transcendental equation which must now be solved by trial and error as in the table of Fig. 2.36 and the plot of Fig. 2.37 from which it can be seen that $\rho_{cr} = 0·20$. The value of P_{cr} is then found from $\rho_1 = P_1/P_{E1}$; or $P_{1cr} = 0·20 P_{E1}$. It is a characteristic of such equations that more than one solution is possible, each solution representing one possible mode. The only mode of real significance to the structural engineer is of course the lowest one.

A corollary to this is that care must be taken to ensure that the solution with the lowest value of axial load parameter is found, otherwise an unsafe estimate of load-carrying capacity will result.

Stability functions may also be used to solve beam-column problems. For example consider again the pin-ended beam-column in Fig. 2.8 which is subjected to a lateral load of W at its midlength.

The distorted shape of this column can be considered to be made up of two distinct deformations, the first being a unit movement δ of B perpendicular to

ρ_1	0·15	0·20	0·25
ρ_2	0·60	0·80	1·00
$s_1(1 + c_1)/m_1$	5·11	4·81	4·52
$2s_1(1 + c_1)/m_1$	10·22	9·63	9·03
s_1	3·80	3·73	3·66
c_1	0·54	0·56	0·57
$(1 + c_1)$	1·54	1·56	1·57
$(1 + c_1)^2$	2·37	2·42	2·47
$s_1(1 + c_1)^2$	9·01	9·02	9·03
$s_2(1 + c_2)/m_2$	2·42	1·21	0
$1·2s_2(1 + c_2)/m_2$	2·90	1·46	0
s_1''	2·69	2·58	2·47
c_2	0·71	0·83	1·00
$(1 + c_2)$	1·71	1·83	2·00
$s_2(1 + c_2)$	5·38	5·16	4·93
$1·5s_2(1 + c_2)$	8·07	7·74	7·40
$\{s_1'' - 1·5s_2(1 + c_2)\}^2$	28·97	26·65	26·04
s_2	3·14	2·82	2·47
$3·75s_2$	11·78	10·56	9·25
$s_1'' + 3·75s_2$	14·47	13·14	11·72
4th term	2·00	2·03	2·80
$H_B \propto$	2·11	0·04	−2·08

Fig. 2.36

the longitudinal axis of the column as indicated in Fig. 2.38. The moments at A, B and C are indicated below together with the magnitude of the lateral force F necessary to cause this distortion.

$$M_{AB} = M_{BA} = -M_{BC} = -M_{CB} = -\frac{EIs(1 + c)\delta}{(L/2)^2} = -\frac{4EIs(1 + c)\delta}{L^2}$$

$$F = \frac{4EIs(1 + c)\delta}{m(L/2)^3} - \frac{32EIs(1 + c)\delta}{mL^3}$$

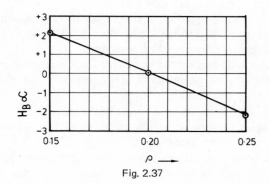

Fig. 2.37

Sway B through δ_B

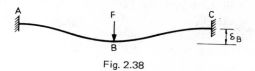

Fig. 2.38

Rotate A through θ and B through $-\theta$

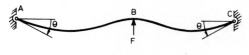

Fig. 2.39

In this distortion pattern there are moments occurring at A and C. This is clearly inadmissible since these ends are pinned. Due to symmetry each end will rotate through an equal amount, θ, but in opposite directions as indicated in Fig. 2.39. The moments arising from such a distortion are listed below together with the magnitude of the lateral force necessary for equilibrium.

$$M_{AB} = -M_{CB} = \frac{EIs\theta}{(L/2)} = \frac{2EIs\theta}{L}$$

$$M_{BA} = -M_{BC} = \frac{EIsc\theta}{(L/2)} = \frac{2EIsc\theta}{L}$$

$$F = \frac{4EIs(1+c)\theta}{L(L/2)} = \frac{8EIs(1+c)\theta}{L^2}$$

To determine the lateral movement produced by the lateral force it is necessary first to eliminate the out-of-balance moments at A and C and the two patterns of distortion must be combined to satisfy the equation

$$M_A = -M_C = -\frac{4EIs(1+c)\delta}{L^2} + \frac{2EIs\theta}{L} = 0$$

giving

$$\theta = 2(1+c)\delta/L$$

Hence the value of the lateral force necessary to cause the distortion δ is given by

$$F = \frac{32EIs(1+c)\delta}{mL^3} - \frac{16EIs(1+c)^2\delta}{L^3}$$

But this force is W, and rearranging, δ is given by

$$\delta = \frac{WL^3}{48EI} \left[\frac{3}{\dfrac{2s(1+c)}{m} - s(1+c)^2} \right] \tag{2.54}$$

Once again this is the normal linear elastic deformation multiplied by an amplification factor in the square brackets. This amplification term when evaluated is identical to that derived from classical analysis. This example demonstrates the versatility of stability functions which are eminently suitable for hand computations now that electronic hand calculators are so readily available. Note that the common factor $s(1 + c)$ has been left in the individual terms. This is because $s(1 + c)/m$ is listed in the tables of Livesley and Chandler under the heading A.

2.15 Bracing

The load-carrying capacity of a slender strut may be increased to any desired level (subject to an upper limit of the squash load) by the provision of bracing, the effect of which is to force the strut into a different buckling shape. Current practice follows what is known as the 'two and a half per cent' rule under which bracing is designed to carry a load of 2½% of the axial load in the braced member. This is a strength criterion but it may be shown that the stiffness of the bracing is also of paramount importance.

Consider the perfect pin-ended strut of Fig. 2.40 which is provided with a single central brace of stiffness K_b, which is defined by $K_b = F/\delta$ where F is the force induced in the brace due to δ, a lateral motion of the strut. Clearly if K_b is small the critical load of the braced strut will only marginally exceed its Euler Load. However, if K_b is large so that the brace acts as a rigid support, as indicated in Fig. 2.41, the strut is forced to buckle in two half waves thereby increasing the critical load to four times that of the unbraced member. Although this problem may be investigated using the basic approach of deriving and solving the governing differential equations, it may also be examined using stability functions and provides a further illustration of their power.

Figures 2.42 (a) and (b) show the two component distortions which when combined in the correct proportions give the symmetrical mode shape of Fig. 2.43. The component distortions consist of a lateral motion of B and equal and opposite rotations of A and C. Also shown are the corresponding moments and the external forces F^1 (additional to the force, F, provided by the brace) which are necessary for equilibrium. A suitable test for instability is the application of equal and opposite rotations of A and C. This will also necessitate ensuring that equilibrium at B is maintained with no external force, i.e. $F^1 = 0$. This is

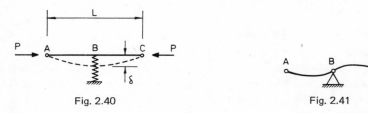

Fig. 2.40   Fig. 2.41

(a)

$$M_{AB} = M_{BA} = -4EIs(1 + c)\delta/L^2$$
$$M_{BC} = M_{CB} = 4EIs(1 + c)\delta/L^2$$
$$F' = -32EIs(1 + c)\delta/(mL^3) - K_b\delta$$

Fig. 2.42

(b)

$$M_{AB} = -M_{CB} = 2EIs\theta/L$$
$$M_{BA} = -M_{BC} = 2EIsc\theta/L$$
$$F' = 8EIs(1 + c)\theta/L^2$$

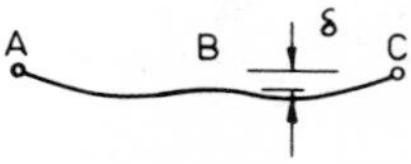

Fig. 2.43

accomplished by taking

$$\left\{ \frac{32EIs(1 + c)}{mL^3} + K_b \right\} \text{ parts of pattern (b) and } -\left\{ \frac{8EIs(1 + c)}{L^3} \right\} \text{ parts of pattern (a)}$$

giving

$$M_{AB} = \left[\left\{ \frac{2EIs}{L} \right\} \left\{ \frac{32EIs(1 + c)}{mL^3} + K_b \right\} - \left\{ \frac{8EIs(1 + c)}{L^2} \right\} \left\{ \frac{4EIs(1 + c)}{L^2} \right\} \right] \theta$$

This moment is a measure of the stiffness of the strut to the imposed distortion and at instability this expression vanishes.

Putting $\mu = K_b L/P_E = K_b L^3/\pi^2 EI$ the instability criterion becomes

$$\frac{32(EI)^2}{L^4} \left\{ s\left[\frac{2s(1 + c)}{m} + \frac{\pi^2\mu}{16} \right] - \left[s(1 + c) \right]^2 \right\} = 0 \qquad (2.55)$$

This must be solved by trial and error for specific values of μ. The results of such computations are shown in the plot of Fig. 2.44 from which it can be seen that the critical load corresponding to the symmetrical mode (as determined from the analysis above) rises markedly as the stiffness of the bracing is increased. For values of μ greater than 16 the critical load for symmetrical buckling becomes larger than the critical load corresponding to buckling into two half waves. Clearly buckling will occur in the mode associated with the lowest critical load. This indicates that once μ exceeds 16 the strut will buckle in an antisymmetrical manner with no deflection of point B and further increase in the stiffness of the brace will no longer produce any increase in critical load. Thus if intermediate braces possess a certain minimum stiffness μ (μ being 16 for the current examples) then they will act as if rigid.

The foregoing analysis relates to a perfect member and yields no information about the force induced in the brace; the analysis merely gives the load at which deflections become uncontrolled. No indications of deformations at intermediate loads are given. Therefore in order to investigate bracing forces it is necessary to perform the analysis on a geometrically imperfect strut. This leads to a more complex analytical problem but it is possible to conduct a simplified investigation using an approximate method first suggested by Winter.

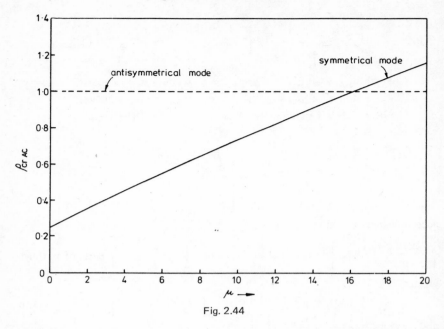

Fig. 2.44

Consider the column of Fig. 2.45 which has an initial lack of straightness defined by a_0 and supported by bracing at midheight. The bracing is required to suppress the lowest mode of instability so that elastic buckling will occur in the mode indicated by the dotted lines — essentially two half sine waves. In this mode $P = \pi^2 \, EI/(L/2)^2$ and there will be a point of inflection at the brace location, i.e. a point of zero moment and thus a pin may be inserted at this point with negligible error. Taking moments about this hinge for equilibrium of one half of the system

$$\frac{FL}{4} - P(a_0 + \delta) = 0 \tag{2.56}$$

But assuming no initial lack of fit of the brace $F = K_b \delta$

$$\frac{K_b \delta L}{4} - P(a_0 + \delta) = 0$$

or

$$K_b = \frac{4P(a_0 + \delta)}{\delta L} = \frac{4P}{L}\left(\frac{a_0}{\delta} + 1\right).$$

Note that for a perfect column $a_0 = 0$ and thus $K_b = 4P/L = 16\pi^2 EI/L^3$ as obtained before. It can however be seen that for an imperfect column ($a_0 \neq 0$) the stiffness required rises with increasing magnitude of initial crookedness. To evaluate this increase it is necessary to know the maximum value of the ratio a_0/δ; this requires that actual values be provided for both the initial imperfection a_0 and the additional deformation δ which may be tolerated. Suppose a_0

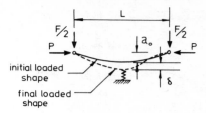

Fig. 2.45

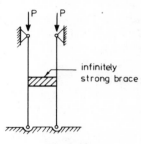

Fig. 2.46

and δ are both taken as $L/500$ then

$$K_{b \, required} = (16\pi^2 EI/L^2)(1 + 1) = 32\pi^2 EI/L^2 \qquad (2.57)$$

which represents a 100% increase in stiffness requirement as compared with the case of the ideal member.

The strength requirement for the bracing may be determined from Eq. 2.56 as

$$F_{required} = 4P\left(\frac{a_0}{L} + \frac{\delta}{L}\right) = 4P\left(\frac{1}{500} + \frac{1}{500}\right) = \frac{8P}{500} \qquad (2.58)$$

or 1.6% of the force in the member being braced.

The current 2½% rule can be rationalised by this simple analysis but note that in this simple rule no mention is made of the minimum stiffness which must also be provided in order that the brace can function as if rigid and completely suppress a mode of instability. As this has only rarely caused problems in the past it would seem that the basic engineering sense of designers has filled the omission in the code rules. It is however pertinent to reiterate the fact that even a brace of infinite strength will not aid the stability of even the most slender member if it is associated with zero stiffness, e.g. the massive connection between the midpoints of the two identical struts of Fig. 2.46.

2.16 Bibliography

1. Bleich, F. (1952) *Buckling Strength of Metal Structures*, New York, McGraw-Hill, ch. 1.

2. Timoshenko, S. P., and Gere, J. M. (1961), *Theory of Elastic Stability*, 2nd edn., New York, McGraw-Hill, chs. 1, 2 and 4.

3. Galambos, T. V. (1968), *Structural Members and Frames*, Englewood Cliffs, N.J., Prentice-Hall, chs. 4 and 5.

4. Trahair, N. S. (1977), *The Behaviour and Design of Steel Structures*, London, Chapman and Hall, chs. 3 and 7.

5. Johnston, B. G. (1977), *Guide to Design Criteria for Metal Compression Members*, 3rd edn., SSRC, New York, Wiley, chs. 3 and 8.

6. Livesley, R. K., and Chandler, D. B. (1956), *Stability Functions for Structural Frameworks*, Manchester University Press.

7. British Standard 449, Part 2: (1969), *Specification for the Use of Structural Steel in Building*, London, BSI.

8. British Standard 153, Parts 3B and 4: (1972), *Specification for Steel Girder Bridges*, London, BSI.

9. *Draft Standard Specification for the Structural Use of Steelwork in Building, Part 1: Simple Construction and Continuous Construction* (1977), London, BSI.

10. *Recommendations for Steel Constructions* (1976), European Convention for Constructional Steelwork, Rotterdam.

CHAPTER THREE

Lateral Buckling of Beams

3.1 Introduction

Beams are normally regarded as structural members whose primary function is the transfer of loads by means of bending action, and in many situations the structural framing is so arranged that the resulting bending may be regarded as being effectively uniaxial, i.e. bending in one plane only. Moreover, by ensuring that the loading acts through the shear centre of the section (which coincides with the centroid for a doubly symmetrical section) twisting will be eliminated. In such cases major-axis bending strength becomes the principal design consideration with the result that the type of cross-section normally selected, e.g. I, channel etc. is usually relatively weak in both minor-axis bending and twisting. This introduces the possibility of lateral or as it is sometimes called lateral-torsional buckling, in which collapse is initiated as a result of lateral deflection

Fig. 3.1 Failure of beam by lateral buckling (*Courtesy of Professor Y. Fukumoto*)

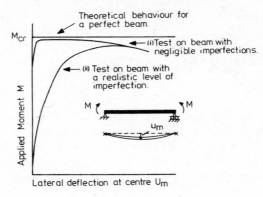

Fig. 3.2

and twisting, as shown in Fig. 3.1. As in the case of struts the exact manner in which these buckling deformations appear depends upon both the initial shape of the beam and the way in which the loading is applied. Typical load versus lateral deflection plots for a slender beam tested under (i) carefully controlled laboratory conditions, and (ii) something approaching service conditions, are shown in Fig. 3.2.

Because the buckling deformations are coupled, the analysis of this class of problem is inherently more complex than that presented in Chapter 2 for pure flexural buckling of struts. Therefore in deriving design rules for beams based on considerations of lateral buckling certain simplifications are essential.

3.2 Twisting of I-Sections

When the beam shown in Fig. 3.3 is twisted about its longitudinal axis then, provided the axial movements required by the deformed shape of Fig. 3.4 are able to occur, it is said to be in pure torsion and the only stresses produced will be shearing stresses distributed approximately as indicated in Fig. 3.5. For a thin-walled open section (such as an I or channel) it may be assumed, with

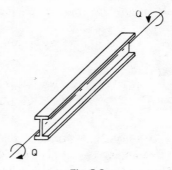

Fig. 3.3

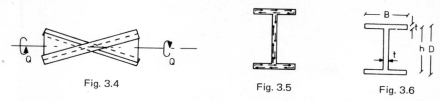

Fig. 3.4   Fig. 3.5   Fig. 3.6

reasonable accuracy, that for any component plate (e.g. web or flange) these shear stresses will act parallel to the edge of that component plate and that their magnitude will be proportional to the distance from the mid-surface of that plate. The torsional deformation, which is conveniently expressed as the angle of twist per unit length ϕ/L, is related to the applied torque Q by

$$\phi/L = Q/C \tag{3.1}$$

where

      C, the torsional rigidity, may be expressed as the product of
      G the shear modulus of the material and
      J the torsion constant for the section.

For a thin-walled open section of constant thickness t, e.g. the I-section shown in Fig. 3.6, the value of J is given approximately by

$$J = \tfrac{1}{3}bt^3 \tag{3.2}$$

where b = length of middle surface of cross-section (e.g. b = 2B + h for the I-section shown in Fig. 3.6). When the component plates have different thicknesses Eq. 3.2 should be replaced by the more general

$$J = \Sigma \tfrac{1}{3}b_i t_i^3 \tag{3.3}$$

where b_i and t_i now relate to plate i and the summation extends over all portions of the cross-section.

Example 3.1

Calculate the value of J for the I-section shown in Fig. 3.6 assuming that

(i)   B = 100 mm, D = 200 mm and web and flange thickness = 10 mm.
(ii) The web thickness is reduced to half that of the flange.

(i) $J = \tfrac{1}{3}[100 \times 10^3 + (200 - 10)10^3 + 100 \times 10^3]$ mm^4
     $= 130\,000$ mm^4

(ii) $= \tfrac{1}{3}[100 \times 10^3 + (200 - 10)5^3 + 100 \times 10^3]$ mm^4
     $= 74\,580$ mm^4

More accurate formulae, which make due allowance for junction effects as well as for the radii and fillets that are present in rolled sections, suggest that Eq. 3.3 will normally give results that are accurate to within a few per cent.

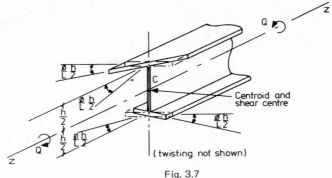

Fig. 3.7

Application of a torque to a section causes the initially straight longitudinal fibres to twist in the form of helices, which, for small angles of twist, can be considered as straight lines inclined to the axis of rotation. Thus for an I-section the central fibres of the flanges, which are at a distance $h/2$ from the z-axis, become inclined to the axis by an angle $\phi h/2L$, thereby inducing axial displacement of the flanges as shown in Fig. 3.7. This type of deformation is called warping and the axial displacements produced are known as the warping displacements. Their appearance means that cross-sections which were initially plane before twisting do not remain plane. For a symmetrical I-section warping may be interpreted as bending of the flanges in opposite directions about a vertical axis parallel to the web. Consideration of the warping of other types of cross-section, e.g. channels and unsymmetrical I-sections, is complicated by the non-coincidence of the centroid of the cross-section and its shear centre.

In discussing the type of behaviour illustrated in Fig. 3.4 it was assumed that the support arrangements and the manner in which the load was applied were both such that axial (warping) deformation was completely free to occur. Thus warping was the same for each cross-section along the length and no axial strains were developed. However, if warping is prevented at any cross-sections, e.g. at the root of the cantilever shown in Fig. 3.8, or is in any way inhibited, e.g. due to adjacent cross-sections wanting to warp by different amounts, then axial stresses will develop in the flanges. In addition, the rate of change of ϕ/L will no longer be constant leading to the expression 'non-uniform torsion' by which warping is sometimes known.

Thus, in the general case, an applied torque will be resisted partly by the development of the shear stresses associated with pure torsion and partly by the development of the axial stresses associated with warping. Denoting these two

Fig. 3.8

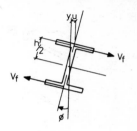

Fig. 3.9

parts as Q_1 and Q_2 respectively, Q_1 is obtained directly from Eq. 3.1 as

$$Q_1 = GJ \, \phi/L$$
$$= GJ \, d\phi/dz \tag{3.4}$$

The second part of the torque, Q_2, is found by considering the bending of the flanges due to warping. Using the symmetry of the section to note that each cross-section will rotate about the z-axis as shown in Fig. 3.9 allows the lateral deflexion of the flanges u to be written as

$$u = \frac{\phi h}{2} \tag{3.5}$$

The lateral bending moment in each flange M_f will be

$$M_f = -EI_f \frac{d^2 u}{dz^2} \tag{3.6}$$

where I_f is the second moment of area of the flange about the y-axis. Substituting for u gives

$$M_f = -EI_f \frac{h}{2} \frac{d^2 \phi}{dz^2} \tag{3.7}$$

The shear force across the width of each flange V_f will be given by

$$V_f = \frac{dM_f}{dz} \tag{3.8}$$

$$= -EI_f \frac{h}{2} \frac{d^3 \phi}{dz^3} \tag{3.9}$$

It is the couple produced by these two shear forces that provides the second part of the resistance to twist Q_2. Thus

$$Q_2 = -EI_f \frac{h^2}{2} \frac{d^3 \phi}{dz^3} \tag{3.10}$$

Combining Eqs. 3.4 and 3.10 gives the complete expression for nonuniform torsion of a symmetrical I-section as

$$Q = GJ \frac{d\phi}{dz} - EI_f \frac{h^2}{2} \frac{d^3\phi}{dz^3} \qquad (3.11)$$

Noting that I_f may be taken as $I_y/2$ and introducing I_w the warping constant enables Eq. 3.11 to be written as

$$Q = GJ \frac{d\phi}{dz} - EI_w \frac{d^3\phi}{dz^3} \qquad (3.12)$$

where $I_w = \frac{h^2}{4} I_y$

Equation 3.12 is also valid for several other forms of cross-section provided the appropriate values of J and I_w are used. Clearly for sections with negligible warping stiffness, e.g. a narrow rectangular section, the second term of Eq. 3.12 may be omitted.

3.3 Buckling of Beams

The behaviour described in this section is based on the usual assumptions involved in the determination of elastic critical or bifurcation loads, namely

(i)   The beam is initially undistorted (no initial bow or initial twist).
(ii) It behaves elastically (no yielding).
(iii) Loads act solely in the plane of the web (no deliberately applied lateral or torsional loads).
(iv) The beam is initially unstressed (no residual stresses).

Clearly in practice these conditions may only be approximately satisfied. Indeed some will definitely be violated, e.g. rolled sections do contain substantial residual stresses. The effects of variations from the ideal case will be discussed in Section 3.4.

When a beam is bent in its plane of greatest stiffness, i.e. about the major axis, by a gradually increasing load, a situation will be reached whereupon it suddenly deflects sideways and twists. Such behaviour is termed lateral or lateral-torsional buckling and the load at which it occurs is the critical load for lateral instability. Figure 3.10, which shows the buckled shape of part of a

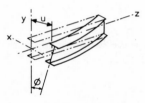

Fig. 3.10

Fig. 3.11

beam, illustrates how lateral buckling involves both a lateral deflexion u of the beam's axis and a twist ϕ about an axis parallel to the z-axis. In order to calculate the magnitude of the critical load it is therefore necessary to consider both actions.

The analysis of lateral buckling based on considerations of equilibrium has much in common with the analysis of the flexural (Euler) buckling of a strut presented in Chapter 2 in that it seeks to find the least load at which the member can exist in its laterally buckled shape.

To illustrate the way in which the analysis proceeds consider the case shown in Fig. 3.11 of a beam loaded by equal and opposite major-axis moments M and supported at both ends in such a way that lateral deflection and twist are both prevented whilst no resistance is provided to lateral bending or warping, i.e. simply supported in the lateral plane. In order to simplify the analysis it will further be assumed that the beam is of narrow rectangular cross-section for which $I_w = 0$; the more realistic and slightly more complex problem of an I-section will be considered later. (Strictly speaking if $I_w = 0$ assumptions regarding the conditions of warping restraint at the ends are unnecessary.) Defining the new coordinate system $\xi\eta\zeta$ shown in Fig. 3.12 which deflects with the beam, the equations governing bending in the two planes $\xi\zeta$ and $\eta\zeta$ are

$$EI_x \frac{d^2 v}{dz^2} = M_\xi \tag{3.13}$$

$$EI_y \frac{d^2 u}{dz^2} = M_\eta \tag{3.14}$$

where M_ξ and M_η are the components of the applied moment M in the $\xi\zeta$ and $\eta\zeta$ planes.

Equations 3.13 and 3.14 assume that ϕ is sufficiently small that the curvatures and flexural rigidities in the $\xi\zeta$ and $\eta\zeta$ planes may be replaced by their values in the xz and yz planes.

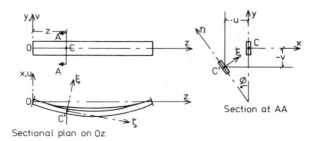

Sectional plan on Oz

Section at AA

Fig. 3.12

The governing equation for twisting is

$$GJ \frac{d\phi}{dz} = M_\zeta \tag{3.15}$$

where M_ζ is the component of M causing twisting about the ζ axis.

Figure 3.13 shows the relationship between the applied moment M, which lies in the yz plane, and its components about the deflected $\xi\eta\zeta$ axis system. Resolution of M into its components about the η and ξ axis as shown in the planar view of Fig. 3.13(b) enables Eqs. 3.13 and 3.14 to be written as

$$EI_x \frac{d^2v}{dz^2} = M \cos \phi \tag{3.16}$$

$$EI_y \frac{d^2u}{dz^2} = M \sin \phi \tag{3.17}$$

which, since ϕ is small, giving $\sin \phi \simeq \phi$ and $\cos \phi \simeq 1$, simplify to

$$EI_x \frac{d^2v}{dz^2} = M \tag{3.18}$$

$$EI_y \frac{d^2u}{dz^2} = M\phi \tag{3.19}$$

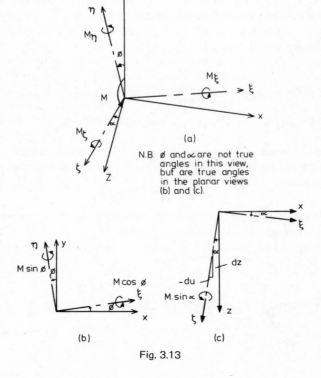

(a)

N.B. ø and ∝ are not true angles in this view, but are true angles in the planar views (b) and (c).

(b)   (c)

Fig. 3.13

The quantity M_ζ may be obtained by noting from the plan view of the xz plane given as Fig. 3.13(c) that it is equal to M sin α. Approximating sin α by $-du/dz$ as shown allows Eq. 3.15 to be written as

$$GJ \frac{d\phi}{dz} = -M \frac{du}{dz}$$ (3.20)

Equations 3.18–3.20 form a system describing the response of the beam to the applied loading M. Inspection reveals that Eq. 3.18 is independent of the other two and, furthermore, that it is the familiar equation of simple in-plane flexure; it has no connection with the beam's lateral buckling behaviour. For the analogous Euler strut problem its counterpart would be the expression relating axial load and axial deformation prior to buckling. On the other hand Eqs. 3.19 and 3.20 are coupled since both contain the deformations u and ϕ. It is these two equations which describe lateral buckling. They may be combined by differentiating Eq. 3.20 and substituting for d^2u/dz^2 in Eq. 3.19 to yield

$$\frac{d^2\phi}{dz^2} + \frac{M^2}{EI_y GJ}\phi = 0$$ (3.21)

This is completely analogous to Eq. 2.3 for the pin-ended column. Therefore following the same solution process and adopting the same arguments gives the critical moment as

$$M_{cr} = \frac{\pi}{L}\sqrt{(EI_y GJ)}$$ (3.22)

The coupled nature of the deformations involved in the buckled shape is clearly reflected in Eq. 3.22 by the presence of EI_y, the beam's lateral bending stiffness, and GJ, its torsional stiffness. The type of load versus twist (or lateral deflection) behaviour implied by this solution is the same as that given previously in Fig. 2.3 for struts, i.e. no buckling deformations for M less than M_{cr} and uncontrolled deformations at $M = M_{cr}$. Solutions based on more complex theory show that for most beams the postbuckling behaviour follows a similar pattern to that of struts; it is therefore normally neglected.

Example 3.11

Determine the value of M_{cr} for a thin flat steel strip of width 12 mm and depth 200 mm when it is used as a beam to span twenty times its depth. What will be the maximum stress in the beam when it is on the point of buckling? Take $E = 205\,000$ N/mm² and $G = 82\,000$ N/mm².

$I_y = 200 \times 12^3/12$ mm⁴

$\quad = 28\,800$ mm⁴

$I_x = 12 \times 200^3/12$ mm⁴

$\quad = 8 \times 10^6$ mm⁴

$$J = \frac{1}{3} \times 200 \times 12^3 \text{ mm}^4$$

$$= 115\ 200 \text{ mm}^4$$

$$M_{cr} = \frac{\pi}{4000} \sqrt{(0 \cdot 205 \times 10^6 \times 28\ 800 \times 0 \cdot 082 \times 10^6 \times 11\ 520)} \text{ Nmm}$$

$$= 1 \cdot 86 \times 10^6 \text{ Nmm}$$

$$p_{cr} = 1 \cdot 86 \times 10^6 \times 100/8 \times 10^6 \text{ N/mm}^2$$

$$= 23 \cdot 2 \text{ N/mm}^2$$

Clearly this stress is much less than the material yield stress and this beam would fail by elastic buckling at a load of approximately one-tenth of that required to initiate yield based on considerations of in-plane bending only.

The foregoing analysis may be extended to cover I-sections simply by ensuring that the fundamental equations (Eqs. 3.13–3.15) are modified appropriately. In fact only Eq. 3.15 need be altered, the inclusion of warping effects giving

$$GJ\frac{d\phi}{dz} - EI_w \frac{d^3\phi}{dz^3} = M_\zeta \qquad (3.23)$$

Substituting for M_ζ and combining with Eq. 3.19 gives

$$EI_w \frac{d^4\phi}{dz^4} - GJ \frac{d^2\phi}{dz^2} + \frac{M^2}{EI_y}\phi = 0 \qquad (3.24)$$

and the solution becomes

$$M_{cr} = \frac{\pi}{L} \sqrt{(EI_y GJ)} \bigg/ \left(1 + \frac{\pi^2 EI_w}{L^2 GJ}\right) \qquad (3.25)$$

The magnitude of the second square root in Eq. 3.25 is a direct measure of the contribution of warping to the torsional resistance of the beam. In the case of a section of low warping stiffness its value approaches unity and, in the limiting case of $I_w = 0$, Eq. 3.25 reduces to Eq. 3.22. On the other hand, an I-section composed of very thin plates will possess a very low torsional rigidity (since J depends upon the third power of the thickness, see Eq. 3.3), and both terms under the root will therefore be of comparable magnitude as shown in Fig. 3.14. In this figure both sections have the same area with the web and flanges of the larger being half the thickness of those of the smaller. If a third section were added comprising the thicker flanges and the thinner web, the results would appear slightly above the lower curve. However, if the abscissa were changed from L/D to L/B (so that the same value of L was used for both narrow flange sections) then the results for this third section would plot much closer to those of the larger section. Thus warping effects will normally be important for sections consisting of thin plates and/or deep webs, becoming relatively insignificant for shallow, stocky sections. Length too has an effect; warping becomes progressively more significant as the beam becomes shorter.

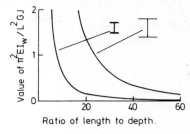

Fig. 3.14

Example 3.III

What would be the effect on M_{cr} of using half the material from the beam of Example 3.II as flanges of width 75 mm and thickness 8 mm with the remainder being left as a web of depth 200 mm and thickness 6 mm?

$$I_y = \frac{1}{12}(2 \times 75^3 \times 8 + 6^3 \times 200)\ \text{mm}^4$$

$$= 566\ 100\ \text{mm}^4$$

$$I_x = \frac{1}{12}(75 \times 216^3 - 69 \times 200^3)\ \text{mm}^4$$

$$= 16.99 \times 10^6\ \text{mm}^4$$

$$J = \frac{1}{3}(2 \times 75 \times 8^3 + 208 \times 6^3)\ \text{mm}^4$$

$$= 40\ 580\ \text{mm}^4$$

$$I_w = 566\ 100 \times 208^2/4\ \text{mm}^6$$

$$= 6123 \times 10^6\ \text{mm}^6$$

$$M_{cr} = \frac{\pi}{4000}\sqrt{(0.205 \times 10^6 \times 566\ 100 \times 0.082 \times 10^6 \times 40\ 580)}$$

$$\times \sqrt{\left(1 + \frac{\pi^2 \times 0.205 \times 10^6 \times 6123 \times 10^6}{4000^2 \times 0.082 \times 10^6 \times 40\ 580}\right)}$$

$$= 17.1 \times 10^6\ \text{Nmm}$$

$$p_{cr} = (17.1 \times 10^6 \times 108)/(16.99 \times 10^6)\ \text{N/mm}^2$$

$$= 109\ \text{N/mm}^2$$

Thus by using the same amount of material in a more efficient fashion the beam's moment-carrying capacity is increased nearly tenfold, although its collapse is still governed by elastic instability. This increase is almost entirely due to the increase in I_y caused by the presence of the flanges since the value of J is actually decreased (because the component plates are thinner) and warping effects are quite small (the value of the second root is only 1.112).

In developing the theory which led to Eq. 3.25, it was assumed that deflections in the plane of the major axis had no effect on lateral torsional behaviour. Including this effect in the formulation of the problem leads to a solution that may be expressed approximately as

$$M_{cr} = \frac{\pi}{L} \sqrt{\left(\frac{EI_y GJ}{\gamma} \right)} \sqrt{\left(1 + \frac{\pi^2 EI_w}{L^2 GJ} \right)}$$

where

$$\gamma = (I_x - I_y)/I_x \tag{3.26}$$

Inspection of Eq. 3.26 shows that as the value of I_y approaches that of I_x so γ tends to 0 and the value of M_{cr} approaches infinity, and that for the special case where I_y exceeds I_x no solution exists. Thus lateral instability is only possible if

(i) the section possesses different stiffnesses in the two principal planes;
(ii) the applied loading causes bending in the stiffer plane.

In practice the effect of major-axis curvature is slight for narrow flange I-sections, e.g. UBs for which I_x is frequently at least twenty times as large as I_y. It can become significant for wide flange H-sections, e.g. UCs, although the relatively large lateral stiffness of such sections means that, when used as beams, their design is unlikely to be greatly influenced by considerations of elastic lateral stability. Since the effect is beneficial it is always safe to neglect it. Indeed in cases where pre-cambering is used it is, of course, essential that it be omitted.

Example 3.IV

What would be the effect on M_{cr} for the beam in Example 3.III of allowing for in-plane deflection?

$$\gamma = (16 \cdot 99 - 0 \cdot 5661) \times 10^6 / 16 \cdot 99 \times 10^6$$
$$= 0 \cdot 967$$

and

$$M_{cr} = 17 \cdot 1 \times 10^6 / \sqrt{(0 \cdot 967)} \text{ Nmm}$$
$$= 17 \cdot 4 \text{ Nmm}$$

i.e. an increase of 1.7%.

The foregoing analyses have suggested that cross-sectional shape is a particularly important parameter in assessing the lateral buckling capacity of a beam. Conversely the problem of lateral instability can be minimised or even eliminated by a judicious choice of section. To illustrate this consider the five different types of section shown drawn to scale in Fig. 3.15(a). Although each has the same cross-sectional area, the values of their flexural and torsional properties relative to those of the unit square exhibit considerable variation. The

Section Properties	▨	▌	╪	I	▭
A	1	1	1	1	1
I_x	1	25	12·45	45·59	16·94
I_y	1	0·04	3·20	3·20	8·10
J	1	0·040	0·034	0·033	4·731

Fig. 3.15(a)

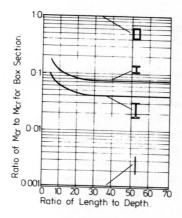

Fig. 3.15(b)

significance of this in the context of lateral stability is illustrated in Fig. 3.15(b), which shows the two sections with the largest in-plane bending stiffnesses (the flat and the deeper I) to be the least stable. Although the very large degree of lateral stability possessed by the box (due principally to its high torsional stiffness) would seem to make it particularly suitable for use as a beam, I-sections are more commonly used because they are easier to produce and more particularly are much easier to join to other members. However, in situations where the beam must be used in a laterally unsupported state, e.g. a crane girder, the possibility of using a box section should be seriously considered.

3.4 Collapse of 'Real' Beams

Clearly the set of assumptions regarding the condition of the beam used as the basis for the theory of the previous Section will not necessarily be satisfied by the types of lateral buckling problem which occur in actual construction. Therefore before the theory can be applied in such cases it is necessary to consider how a failure to comply with any of these assumptions will affect the beam's behaviour.

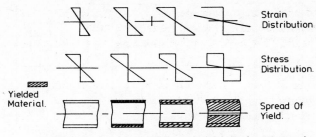

Strain Distribution

Stress Distribution

Yielded Material.

Spread Of Yield.

NB. Elastic - perfectly plastic material behaviour assumed.

Fig. 3.16

Consider first the case where buckling need not necessarily be elastic, i.e. assumption (ii) is relaxed but assumptions (i), (iii) and (iv) are retained. Since, by definition, a beam is a member subjected to loads causing bending, the distribution of axial strain through the depth of the section at any point along the span will be as shown in Fig. 3.16. Thus as the applied moment is increased, yielding will spread gradually through the section, beginning at the outer faces of both flanges. Only in those cases for which the elastic critical moment of Eq. 3.25 is less than the moment at first yield, i.e. $M_{cr} < M_Y$, will lateral buckling be a purely elastic phenomenon. If $M_{cr} > M_Y$ then buckling will not occur until after the appearance of some plastic zones. The limiting case will, of course, correspond to the beam that is sufficiently stocky for it to attain its fully plastic moment M_p. (For plastic design it is, of course, also necessary that M_p be maintained so as to give adequate plastic hinge rotation; this problem is, however, outside the scope of the present volume.) This interaction between instability and plasticity is summarised in Fig. 3.17, on which three distinct regions may be observed:

(a) Beams of high slenderness ($\sqrt{(M_p/M_{cr})} > 1 \cdot 1$) which fail by elastic lateral buckling at M_{cr}

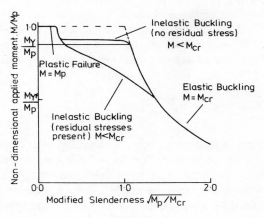

Fig. 3.17

(b) Beams of intermediate slenderness $(1 \cdot 1 > \sqrt{(M_p/M_{cr})} > 0 \cdot 4)$ for which collapse is by inelastic lateral buckling at loads below M_{cr}
(c) Stocky beams $(0 \cdot 4 > \sqrt{(M_p/M_{cr})})$ which are capable of attaining M_P without buckling.

The presence of residual stresses in structural steel members has been discussed previously in Section 2.8, both in terms of the reasons for their presence and with particular reference to their influence on column buckling. For beams their effect is similar in that they cause yielding to be initiated at lower moments and once started yielding spreads more gradually through the cross-section as the applied moment is increased. Thus, as shown in Fig. 3.17, the inelastic range, which now starts at a moment M_{Yr}, is increased at the expense of the elastic range. The presence (or absence) of residual stresses has no effect upon the value of M_P and the plastic range is virtually unaffected.

In order to examine the effect of either the presence of an initial lack of straightness (which, since lateral buckling involves two types of deformation, may be either an initial bow or an initial twist) or of eccentricity in the applied loading, it is convenient to consider first the case where the beam is both free of residual stresses and very slender such that $M_{cr} \ll M_Y$. It has previously been shown for struts that geometrical imperfections and load eccentricity produce qualitatively similar effects; this is also true for beams. Thus for the 'imperfect' beam lateral deflection and twist increase continuously from the start of loading, tending to become very large as the applied moment approaches M_{cr}. These additional deformations produce additional stresses (lateral bending stresses and warping stresses), and for very slender beams failure will occur almost immediately after the maximum stress in the beam (which now occurs at the tip of one flange only) reaches the material yield stresses. A beam 'failure curve' based on this limiting stress type of approach is shown in Fig. 3.18. For beams of intermediate slenderness some redistribution of stress after initial yielding is possible, thereby causing the approach to underestimate the true strength. When the presence of residual stresses is allowed for in calculating the values of the

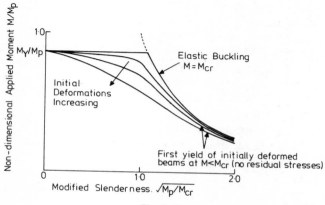

Fig. 3.18

applied loads at which yielding is initiated it is found that the approach becomes conservative even for very slender beams.

In order to understand the behaviour of real beams it is therefore necessary to consider the combined effects of instability and plasticity and, furthermore, to appreciate the role of factors such as residual stresses and geometrical imperfections. When all of these effects are correctly taken into account the resulting situation is found to be one in which very slender beams fail more or less elastically by excessive lateral deformation at loads that are close to M_{cr}, beams of intermediate slenderness fail inelastically by excessive lateral deformation, whilst stocky beams will attain M_P with negligible lateral deformation. A satisfactory design approach for beams must implicitly recognise these different types of failure in presenting a relationship between beam strength and beam geometry. This relationship must balance the conflicting requirements of tolerable accuracy and ease of evaluation.

3.5 Application of Lateral Buckling Theory to the Design of Beams

It is evident from the previous section that considerations of lateral instability will influence the design of all but the stockiest beams. However, because of the extent to which the behaviour of real beams in the practical range of slendernesses is affected by factors not considered in the development of the theory of Section 3.3, only in the case of extremely slender beams could the value of M_{cr} obtained from Eq. 3.25 be expected to provide a suitable basis for design. For beams of intermediate slenderness, where failure is governed by the interaction of both instability and plasticity effects, some form of transition between in-plane collapse and elastic buckling is necessary. Despite this the value of M_{cr} remains of fundamental importance in assessing the load-carrying capacity of a beam, since the ratio of the elastic critical load to the plastic collapse load (as used for example in the non-dimensional slenderness $\sqrt{[M_p/M_{cr}]}$) provides a good indication of the importance of instability and therefore also of the governing mode of failure.

The requirements outlined above may be satisfied in a number of ways, a situation that is reflected in the various methods of dealing with the design of laterally unbraced beams used in different national codes of practice. The approach used in Britain may conveniently be regarded as being composed of two stages:

(i) Provision of a convenient method of evaluating M_{cr}
(ii) Derivation of a suitable design curve giving design strength as a function of M_{cr} and therefore also of beam geometry.

The necessity for the first of these stages is a direct consequence of the complexity of the basic expression for M_{cr} (Eq. 3.25 of Section 3.3), i.e.:

$$M_{cr} = \frac{\pi}{L} \sqrt{(EI_y GJ)} \Big/ \left(1 + \frac{\pi^2 EI_w}{L^2 GJ}\right)$$

It has already been explained in Section 3.3 how for hot-rolled sections and plate girders of 'normal' proportions the two terms under the second root are of comparable magnitude. Therefore only for very long beams (for which the second term is negligible) or for very short beams (for which the first term is negligible) is it permissible to omit part of the solution. Even if this were possible it would, of course, mean that the resulting 'simplified' solutions would be limited in their application. (The situation is rather different for cold-formed sections which, because of their thinness, have very low torsional rigidities GJ, thereby enabling Eq. 3.25 to be simplified without the same loss of generality.)

The 1969 version of BS 449 (as well as the 1972 version of BS 153) uses as its starting point Eq. 3.25 rewritten in terms of the maximum bending stress p_{cr} in the beam when it is on the point of buckling.

$$p_{cr} = \frac{\pi^2 EA}{Z_x \lambda^2} \left(\frac{D-T}{2}\right) \Big/ \left(1 + \frac{4GJ\lambda^2}{\pi^2 EA(D-T)^2}\right) \tag{3.27}$$

where

Z_x = elastic major axis section modulus
A = area of cross-section
$\lambda = L/r_y$ is the slenderness ratio
r_y = minor radius of gyration
D = overall depth
T = flange thickness

Studies conducted by Kerensky, Flint and Brown reported in the Proceedings of the Institution of Civil Engineers in August 1956 showed that for the range of sections then in use an acceptable simplification of Eq. 3.27 could be obtained by substituting the following set of assumed cross-sectional relationships:

$$Z_x \simeq 1 \cdot 1\ BTD \qquad D - T \simeq D \qquad I_y \simeq B^3 T/6$$
$$J \simeq 0 \cdot 9\ BT^3 \qquad A \simeq 2 \cdot 94\ BT \qquad B \simeq 4 \cdot 2\ r_y \tag{3.28}$$

This gives after some additional rounding

$$p_{cr} = \left(\frac{1675}{\lambda}\right)^2 \Big/ \left[1 + \frac{1}{20}\left(\lambda \frac{T}{D}\right)^2\right] \ \text{N/mm}^2 \tag{3.29}$$

Equation 3.29 contains only two geometrical quantities L/r_y and D/T; both are tabulated in section handbooks. Solutions to Eq. 3.29 are given in Table 7 of BS 449. Since the approximations of Eq. 3.28 were at the time of their introduction considered to be generally conservative for 'handbook sections' a 20% increase in p_{cr} is permitted for the majority of cases.

Example 3.V

Compare the value of M_{cr} given by Eq. 3.29 with the 'exact' value as determined from Eq. 3.25 for a 4 m long 457 × 152 UB 52.
From section tables r_y = 3·11 cm, D = 449·8 mm, T = 10·9 mm, Z_x = 949·0 cm³.

$$\therefore \quad \lambda = 400/3 \cdot 11 = 129.$$

For this section p_{cr} given by Eq. 3.29 may be increased by 20%

$$\therefore p_{cr} = 1 \cdot 2 \left(\frac{1675}{129}\right)^2 \Big/ \sqrt{\left[1 + \frac{1}{20}\left(129 \frac{10 \cdot 9}{449 \cdot 9}\right)^2\right]} \; N/mm^2$$

$$= 247 \; N/mm^2$$

and

$$M_{cr} = 247 \times 949 \; kNmm$$

$$= 234 \; kNm$$

From section tables I_y = 645 cm⁴, J = 21·26 cm⁴

$$\therefore \quad I_w = 645\,[(44 \cdot 98 - 1 \cdot 09)^2/4] \; cm^6$$

$$= 310\,600 \; cm^6$$

$$M_{cr} = \frac{\pi}{4000} \sqrt{(0 \cdot 205 \times 10^6 \times 6\,450\,000 \times 0 \cdot 082 \times 10^6 \times 212\,600)}$$

$$\times \sqrt{\left(1 + \frac{\pi^2 \times 0 \cdot 205 \times 10^6 \times 310\,600 \times 10^6}{4000^2 \times 0 \cdot 082 \times 10^6 \times 212\,600}\right)} Nmm$$

$$= 215 \; kNm$$

The code method therefore gives a value that is 8% high in this case. This is largely a result of the approximations of Eq. 3.28, since these were based on an earlier range of standard sections, i.e. not UBs or UCs.

The transition from the critical stress given by Eq. 3.29 to a permissible or design stress in the 1969 version of BS 449 is achieved by using an empirically adjusted Perry-Robertson approach. This limits the maximum stress in an initially bowed (and initially stress-free) beam caused by simultaneous lateral and in-plane bending and warping to the material yield stress p_Y. The actual design curve corresponds to the case of D/T → ∞ with η, the maximum initial bow, being taken as

$$\eta = 0 \cdot 003\lambda \qquad\qquad (3.30)$$

This leads to the following expression for p_b the nominal applied stress

$$p_b^3 - p_b^2(p_Y + 385\lambda) - p_b p_{cr}(p_{cr} + 385\lambda) + p_Y p_{cr}^2 = 0 \qquad (3.31)$$

Clearly below a certain limiting slenderness lateral buckling effects become negligible, and this is reflected in the use of $p_b = p_Y$ for values of λ up to 60. Between values of λ of 60 and 100 a straight-line transition is used, with the results of Eq. 3.31 only being used for higher values of λ. The complete curve is

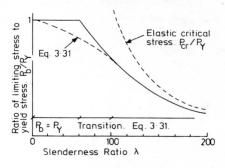

Fig. 3.19

shown in Fig. 3.19, values of permissible design stress p_{bc} being obtained from these values of p_b by dividing by a safety factor whose value varies from about 1·7 at high slendernesses to about 1·52 at low slendernesses. This variation allows for the fact that elastic buckling is an ultimate condition whereas the attainment of first yield in a stocky beam is not (due to the shape factor effect the actual factor on M_p is about 1·7). The code tabulates values of p_{bc} against p_{cr}. An alternative, more direct approach is included in BS 449 (but not BS 153) in which p_{bc} is tabulated directly against λ and D/T for rolled sections, i.e. those which meet the condition for the application of the 'plus 20%' clause on p_{cr}.

Example 3. VI

Assuming Grade 43 steel what is the value of permissible bending stress for the beam of the previous example?

From Table 8 of BS 449 : 1969 for p_{cr} = 248 N/mm², p_{bc} = 87 N/mm².
Alternatively using λ = 129 and D/T = 41·3 in Table 3a directly gives p_{bc} = 128 N/mm². This example shows up certain discrepancies between the 'direct approach' (Table 3a) and the 'two-stage approach' (Table 8). It would appear that this is due to certain increases in the values of Table 3 which occurred as part of an amendment to the original version. Moreover, there is some evidence to suggest that the use of Table 3 leads to rather high results.

Whereas the 1969 version of BS 449 was written in terms of allowable stresses the new version adopts the philosophy of limit states. Therefore since the theoretical maximum moment capacity of a beam that is not susceptible to prior failure by buckling is its fully plastic moment M_p it is logical to express beam strength as the proportion of M_p that can be developed. The new code also employs the idea of a 'lateral-torsional' slenderness λ_{LT} to write the design capacity M_b as

$$\frac{M_b}{M_p} = f\left(\frac{1}{\lambda_{LT}^2}\right) \qquad (3.32)$$

which is analogous to the expression for non-dimensional strut capacity

$$\frac{p}{p_Y} = f\left(\frac{1}{\lambda^2}\right)$$

The quantity λ_{LT} is defined by

$$\lambda_{LT} = \sqrt{\left(\frac{\pi^2 E}{p_Y}\right)} \sqrt{\left(\frac{M_p}{M_{cr}}\right)} \tag{3.33}$$

which for a particular material (and hence a particular value of both E and p_Y) corresponds to the product of a pure number (about 90 for Grade 43 steel depending on the exact values of E and p_Y adopted) and the modified slenderness used previously in Section 3.4 to distinguish between the various types of beam failure.

. Substituting in Eq. 3.33 for both M_p and M_{cr} enables λ_{LT} to be written purely in terms of the beam's geometrical properties as

$$\lambda_{LT} = \frac{u\lambda}{\sqrt[4]{\left[1 + \frac{1}{20}\left(\lambda\frac{1}{x}\right)^2\right]}} \tag{3.34}$$

where

$\quad u = \sqrt[4]{[(S_x/Ah)^2 4\gamma]}$ is the buckling parameter $\tag{3.35}$

$\quad x = 0{\cdot}566 \sqrt{(A/J)}$ is the torsional index $\tag{3.36}$

and

$\quad S_x$ is the plastic section modulus (major-axis).

Studies of the values of u for currently available sections show that it varies between about $0{\cdot}7$ and $1{\cdot}0$, being approximately $0{\cdot}9$ for narrow flange I-sections, e.g. UBs and channels and approximately $0{\cdot}85$ for wide-flange I-sections, e.g. UCs. Furthermore x is approximately equal to D/T for rolled sections. It is expected that values of u and x will eventually be tabulated in section handbooks (at the time of writing tables are provided in the draft code). Alternatively, the following safe approximations given in the code may be used:

For a rolled I, H or channel $u = 0{\cdot}9$
For any other case $u = 1{\cdot}0$
For all sections $x = D/T$

Rapid evaluation of λ_{LT} is facilitated by the use of Eq. 3.34 in the alternative form

$$\lambda_{LT} = uv\lambda \tag{3.37}$$

in which

$$v = \frac{1}{\sqrt[4]{\left[1 + \frac{1}{20}\left(\lambda\frac{1}{x}\right)^2\right]}}$$

may be obtained directly from the table (or graph) given in the code.

Example 3. VII

Use Eq. 3.34 to calculate the value of M_{cr} for a 457 × 152 UB 52 when it is simply supported over a span of 4 m.

From tables $r_y = 3 \cdot 11$ cm, $u = 0 \cdot 859$, $x = 44 \cdot 0$, $S_x = 1094$ cm³

$$\therefore \quad \lambda = 400/3 \cdot 11 = 129$$

$$\lambda_{LT} = \frac{0 \cdot 859 \times 129}{\sqrt[4]{\left[1 + \frac{1}{20} \left(129 \frac{1}{44 \cdot 0} \right)^2 \right]}}$$

$$= 101$$

From Eq. 3.33,

$$M_{cr} = \frac{\pi^2 E}{p_Y} \frac{M_p}{\lambda_{LT}^2}$$

which, noting that $M_p = S_x p_Y$, gives

$$M_{cr} = \pi^2 \times 0 \cdot 205 \times 10^6 \times 1\ 094\ 000/101^2 \text{ N mm}$$

$$= 217 \text{ kNm}$$

which compares very favourably with the 'exact' value obtained in example 3.V.

This stage corresponds to step (i) on page 78, although the introduction of λ_{LT} has, of course, rather altered the appearance of the problem. Nevertheless the change is merely one of presentation. It now remains to use λ_{LT} directly in specifying a design curve for M_b, i.e. to find the form of the function in Eq. 3.32.

Following on from the representation of the European column curves by a modified Perry formula discussed at the end of Section 2.9, beam design in the new code is also based on a modified Perry type of interaction formula (the interaction is, of course, one of elastic instability and plasticity) with M_b being taken as the smaller root of

$$(M_{cr} - M_b)(M_p - M_b) = \eta_{LT} M_b M_{cr} \tag{3.38}$$

in which

$$\eta_{LT} = 0 \cdot 007 \left[\lambda_{LT} - 0 \cdot 4 \sqrt{\left(\frac{\pi^2 E}{p_Y} \right)} \right] \tag{3.39}$$

The resulting design curve is shown (in non-dimensional form) in Fig. 3.20. Solutions to Eq. 3.38 are presented in the new code as a series of tables (and graphs) of design strength $p_b = M_b/S_x$ (where S_x is the plastic section modulus) against λ_{LT}. Therefore a check on the capacity of a trial section using the new procedure involves two steps:

(i) Evaluation of λ_{LT} using u, x and λ – Eq. 3.37.
(ii) Determination of the corresponding value of p_b – Eq. 3.38.

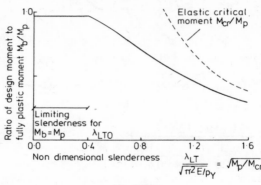

Fig. 3.20

In making this check the correct partial safety factors should be used at the appropriate stages.

Example 3. VIII

Assuming steel of yield strength $p_Y = 240\ N/mm^2$, what is the ultimate moment capacity of the beam of the previous example?

From Table 6.2.1 of the draft code for $\lambda_{LT} = 101$ and $p_Y = 240\ N/mm^2$

$p_b = 116\ N/mm^2$

∴   moment capacity $M_b = p_b S_x$

$$= 116 \times 1094\ Nm$$

$$= 127\ kNm$$

which, noting that $M_p = 263\ kNm$, is approximately one-half of its capacity when fully braced. The value of M_b calculated above should, of course, be compared with the factored loads.

3.6 Effect of Non-Uniform Moment

So far only the case of beams loaded with equal and opposite end moments, i.e. uniform single curvature bending, has been considered. Beams in practical situations will, of course, be subject to a whole range of different loading conditions which will in turn produce a variety of different patterns of moment and some examples are shown in Fig. 3.21. Indeed, the case of uniform bending is rarely encountered in practice. This naturally leads to the question 'Why consider this as the basic case?'. There are two reasons: first it is the easiest to treat analytically and therefore the easiest with which to demonstrate the basic principles and second it is generally regarded as the most severe loading condition. This second reason means, of course, that correct allowance for the

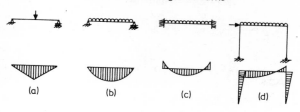

Fig. 3.21

(usually) less severe conditions encountered most frequently in practice will lead to economies in design. The usual approach to the problem is to base the modification to the basic design procedure of the previous section on a comparison of the elastic critical load for the actual case with the elastic critical load for the basic case, a process that has much in common with the use of the effective length concept for dealing with end fixity in strut problems.

A situation that is frequently encountered in practice is that in which a beam is subjected to loads which act only at points of effective lateral restraint, e.g. Example (a) of Fig. 3.21 if the load is applied by a cross-beam. The variation of bending moment within a segment (length of beam between restrained points) will therefore be linear from M at one end to βM at the other end as indicated in the inset to Fig. 3.22. For any value of β the value of M at which elastic instability will occur may be obtained directly from Fig. 3.22 in which it is expressed as a ratio involving the value of M_{cr} given by Eq. 3.25, i.e. the elastic critical moment for $\beta = 1 \cdot 0$. Denoting this ratio by m it is clear that since $m \leqslant 1 \cdot 0$, moment gradient is a less-severe condition, in terms of its effect upon lateral stability, than uniform moment. Clearly the results of Fig. 3.22 may safely be approximated by a single curve and a suitable version is:

$$m = 0 \cdot 57 + 0 \cdot 33\beta + 0 \cdot 1\beta^2 \not< 0 \cdot 43 \tag{3.40}$$

The quantity m is usually referred to as an 'equivalent uniform moment factor'; its value gives a direct measure of the severity of the actual pattern of moments as compared with the basic pattern used in establishing the design curve.

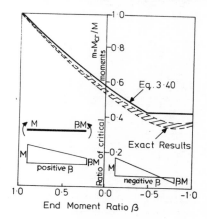

Fig. 3.22

Example 3.IX

What is the value of m for a 457 × 152 UB 52 beam of span 4 m subjected to end moments 300 kNM and 72 kNm? (Both moments act in a clockwise sense.)

From Fig. 3.22 since the end moments are tending to produce double curvature β will be negative.

$$\beta = -72/300$$

$$= -0.24$$

From Eq. 3.40 (or alternatively using Fig. 3.22),

$$m = 0{\cdot}57 + 0{\cdot}33(-0{\cdot}24) + 0{\cdot}1(-0{\cdot}24)^2$$

$$= 0{\cdot}497$$

i.e. this form of loading is approximately half as severe as uniform moment in terms of its effect on the elastic lateral stability of the beam.

It is also possible to relate the critical loads for other loading cases to M_{cr} and hence to derive values of m. In many cases it has been found that m is virtually independent of all factors other than the shape of the moment diagram and some examples are given in Fig. 3.23. For all of these cases very good estimates of the critical moment due to the actual loading may be found using the appropriate value of m in the expression

$$M = \frac{1}{m} M_{cr} \qquad c_b = \frac{1}{m} \tag{3.41}$$

where the value of M_{cr} is obtained from Eq. 3.25.

It is of course possible, in theory, to postulate an infinite variety of possible moment patterns. Clearly not all of these are likely to be encountered in practice

Beam & Loads	Actual bending moment	M_{max}	m	Equivalent uniform moment
		M	1·00	
		M	0·57	
		M	0·43	
		$\frac{WL}{4}$	0·74	
		$\frac{wL^2}{8}$	0·88	
		$\frac{WL}{4}$	0·96	
		$\frac{3WL}{16}$	0·69	
		$\frac{WL}{8}$	0·59	
		$\frac{wL^2}{12}$	0·39	

Fig. 3.23

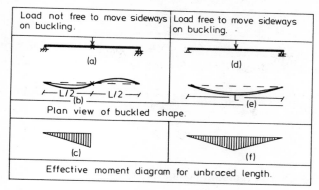

Fig. 3.24

	Moment Pattern	Type of Variation
1		Linear
2		Bilinear
3		Trilinear
4		Parabolic
5		Non-linear
6	Combinations of the above including shifts in the base line due to end fixity.	

Fig. 3.25

so it is necessary to identify the most common cases. In doing this it is important to distinguish between loads applied in such a way that the method of transferring the load automatically provides lateral restraint at the load point, e.g. loads applied by means of cross-beams, and loads which are free to move sideways as the beam buckles, e.g. loads applied by a crane trolley. Taking the particular example of a beam subjected to a concentrated load as shown in Fig. 3.24, if the loading is assumed to be of the first kind the moment pattern will correspond to a $\beta = 0.0$ moment gradient in both segments (Fig. 3.24(c)) the unbraced length will be $L/2$ and the plan view of the buckled shape will contain a node at the load point as shown in Fig. 3.24(b). When the loading is of the second kind the moment pattern will be as shown in Fig. 3.24(f), the unbraced length will now be the full span L, and the plan view of the buckled shape will contain a point of maximum displacement at the load point, Fig. 3.24(e).

When the above comments are borne in mind it will be found that the list of moment patterns given in Fig. 3.25 is adequate for the approximation of most of the cases that are likely to be encountered. Cases 1, 2 and 4 are straightforward and have already been dealt with in Fig. 3.23, whilst a method of dealing with combined end moments and transverse loading is given in the draft code. For moment patterns that cannot readily be approximated to 'fit' one of the

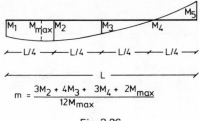

$$m = \frac{3M_2 + 4M_3 + 3M_4 + 2M_{max}}{12M_{max}}$$

Fig. 3.26

standard categories the approach described in Fig. 3.26 may be used to obtain good estimates of the critical moment.

3.7 Use of m-Factors in Design

This section will deal specifically with the proposals given in the new draft code to allow for the shape of the moment diagram. These constitute a major revision of the 1969 version for which the omission of such guidance effectively meant that all beams were designed for uniform moment loading, a process that is unnecessarily conservative in many cases, particularly where slender beams are involved.

The method follows the process common to a number of buckling situations[*] of using the basic curve of Fig. 3.20 in conjunction with a modification based on a comparison of the elastic critical load for the basic case. It is possible to do this in two ways:

(i) Use the value of the equivalent uniform moment $\bar{M} = mM_{max}$ when checking the capacity against M_b

(ii) In calculating M_b from Eq. 3.38 replace M'_{cr} by M_{cr}/m, i.e. enter Fig. 3.20 with an effective value $\lambda^1_{LT} = \lambda_{LT}\sqrt{m}$.

Since M_b will always be less than M_{cr}, considerably so for low values of λ_{LT}, the second method is more conservative than the first as illustrated in Fig. 3.27. (In order that the two methods may be compared on the same figure, method (i) is shown in its 'inverse form', i.e. M_{max} is to be checked against M_b/m.)

Method (i) has been found to be appropriate in all cases of moment gradient loading, i.e. loads applied only at points of effective lateral restraint. This is due to the restriction of yielding to a region near the support(s) resulting in only a relatively small reduction in lateral buckling strength. When used in design this approach requires the additional check that $M_{max} \not> M_b$ in order to prevent prior failure by overstressing at one end.

---

[*]An example would be the use of an effective length in column design as discussed at the end of Section 2.10.

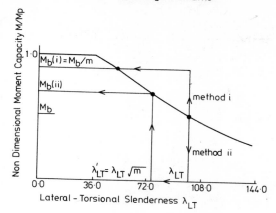

Fig. 3.27

Example 3.X

Check whether the beam of Example 3.VII would be safe under factored-applied loading which produces end moments of 155 kNm (clockwise) and 86 kNm (anticlockwise).

$\beta = 86/155$

$\quad = 0.555$

From Eq. 3.40

$\quad m = 0.57 + 0.33\,(0.555) + 0.1\,(0.555)^2$

$\quad = 0.784$

$\therefore \quad$ equivalent uniform moment $\bar{M} = 0.784 \times 155$ kNm

$\quad\quad\quad\quad\quad\quad = 122$ kNm

From Example 3.8 $\lambda_{LT} = 101$ and $M_b = 127$ kNm
Thus the capacity (127 kNm) exceeds the design moment (122 kNm) and the beam is safe.

In cases where the point of maximum moment occurs within the span the reductions in stiffness due to yield usually exceed the benefits of the less severe pattern of moments and method (i) tends to overestimate the beam's capacity. Method (ii) is therefore appropriate.

It is presented in the code in terms of a series of slenderness correction factors n to be applied directly to the value of λ_{LT} obtained from Eq. 3.37. Thus the design strength should be obtained for a slenderness $\lambda'_{LT} = n\,\lambda_{LT}$; clearly n is simply equal to $\sqrt{m}$.

Example 3.XI

If the loading on the beam of the previous example is changed to a central load which produces a maximum (factored) moment of 155 kNm will the beam still be safe?

From Fig. 3.23, m = 0·74

$$\therefore \quad n = \sqrt{0·74}$$

$$= 0·86$$

and

$$\lambda^1_{LT} = 0·86 \times 101$$

$$= 87$$

From Table 6.2.1 of Draft Code for $\lambda_{LT} = 87$ and $p_Y = 240$ N/mm²

$$p_b = 140 \text{ N/mm}^2$$
$$M_b = 140 \times 1·094 \text{ kNm}$$
$$= 153 \text{ kNm}$$

Thus the capacity (153 kNm) is just less than the design moment (155 kNm) and the beam is not safe. (In practice the designer might accept this since the difference is small.)

This example, together with the previous example, illustrate the differences between the two methods of allowing for the pattern of moments.

3.8 Effect of Level of Application of Transverse Loads

When a beam is subjected to a system of transverse loads its lateral stability is dependent not only on the arrangement of the loads within the span (which affects the pattern of moments) but also on the level of application of such loading relative to the centroid of the cross-section. Figure 3.28, which presents results for the particular case of a centrally-loaded beam, shows how this is due to the additional destabilising or restoring effect which occurs as soon as the loaded cross-section is twisted. As might be expected, the magnitude of this effect is dependent on the proportions of the beam, being most significant in those cases where warping effects are important.* Clearly if the transverse loading is applied in such a way that twisting of the loaded cross-section(s) is prevented then the actual level of application of the loading will have no effect.

Whilst it may be desirable to allow for the beneficial effects of loads applied below the centroid, e.g. a runway beam with the hoist suspended from the bottom flange, it is clearly imperative that the deleterious effects of loads which act above the centroid, e.g. loading applied to the top flange through a light-weight deck which is incapable of providing any lateral restraint, be allowed for if unsafe designs are to be avoided. Whereas the 1969 version of BS 449 considered only the case of top flange loads, the new draft code gives the topic more attention.

The most academically satisfactory way of dealing with this problem would be to follow the procedure of the previous section with the value of n being

*Refer back to Section 3.3 for an explanation of the parameter $L^2 GJ/EI_w$ used in Fig. 3.28.

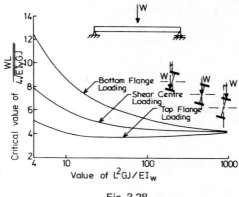

Fig. 3.28

selected so that it reflected both the pattern of moments and the level of application of the loading. Whilst this procedure is permitted by the new code, its use requires a certain familiarity with published data. Therefore, as an alternative the simple 'blanket allowance' of the previous code which requires the use of a notional effective length* of $1 \cdot 2$ times the actual span to be used in the calculation of λ_{LT} has been retained. However, the scope of this approach has been extended somewhat by the inclusion of a pro rata increase for loads which act above the level of the top flange.

Example 3.XII

What would be the value of the effective slenderness λ'_{LT} for the beam in Example 3.VII if the load was applied at the level of the top flange?

Using Table 6.5 on page 264 of Timoshenko and Gere enables the 'exact' value of M_{cr} for this form of loading to be determined as follows:

Value of

$$L^2 GJ/EI_w^{\dagger} = 4000^2 \times 82\,000 \times 212\,600/205\,000 \times 310\,600 \times 10^6$$

$$\therefore \quad \gamma_2 = 19 \cdot 7$$

and

$$M_{cr} = 19 \cdot 7/4 \times 4000 \sqrt{(205\,000 \times 6\,450\,000 \times 82\,000 \times 212\,600)}\ \text{Nmm}$$

$$= 186 \cdot 9\ \text{kNm}$$

from Eq. 3.33

$$\lambda'_{LT} = \pi\sqrt{(205\,000/p_Y)} \sqrt{(1\,094\,000\ p_Y/186 \cdot 9 \times 10^6)}$$

$$= 109$$

---

*For a discussion of the exact meaning of this term see Section 3.9.
†Termed $L^2\ C/C_1$ in Timoshenko and Gere.

Alternatively, using $1{\cdot}2\,\lambda$ in Eq. 3.38 gives

$$\lambda'_{LT} = \frac{0{\cdot}859 \times 1{\cdot}2 \times 129}{\sqrt[4]{\left(1 + \frac{1}{20}\ \ 1{\cdot}2 \times 129 \times \frac{1}{44{\cdot}0}\right)^2}}$$

$$= 118$$

Thus the approximate method gives results which are more conservative than the true value.

3.9 Effect of End Support Condition

The support conditions assumed in the derivation of the basic theory of Section 3.3, and used in all the succeeding discussions, are illustrated schematically in Fig. 3.29. This situation would be approximated in practice by any arrangement in which restraint was supplied only to the web of the beam, e.g. by means of web cleats in the case of a beam supported by a column as illustrated in Fig. 3.30. The important requirement is that the connection be capable of preventing both lateral deflection and twisting.

Clearly it is necessary to be able to assess to what extent the various other forms of support that arise in practice will affect the beam's lateral stability. For the present only those beams which consist of a single span in both the transverse and the lateral planes will be considered; the behaviour of continuous beams will be discussed separately in a later section.

Studies of the influence of end restraint on the elastic buckling of beams loaded by uniform moment have shown that the results may conveniently be expressed in the form

$$M_{cr} = \frac{\pi}{k_1 L} \sqrt{(EI_y GJ)} \ \sqrt{\left(1 + \frac{\pi^2 EI_w}{k_2^2 L^2 GJ}\right)} \tag{3.42}$$

in which k_1 and k_2 are effective length factors.

The presence of two k factors in Eq. 3.42 reflects the two possible types of end fixity, lateral bending restraint and warping restraint. Values of k_1 and k_2 for three standard cases are given in Fig. 3.31. For other forms of applied loading the problem is complicated by the fact that k_1 and k_2 are frequently not even approximately constant but vary with the proportions of the beam.

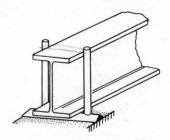

Fig. 3.29

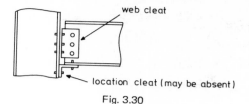

web cleat

location cleat (may be absent)

Fig. 3.30

Type of end conditions	k_1	k_2
Simply Supported	1·00	1·00
Warping Fixed	0·92*	0·48*
Completely Fixed	0·50	0·50
*Approximate value		

Fig. 3.31

Accurate assessment of the degree of restraint provided by practical forms of connection is difficult. A condition of full end fixity (in the lateral plane) will only be approached by arrangements which involve complete continuity at the joint, e.g. the beam-to-column connection shown in Fig. 3.32, although any arrangement which provides positive attachment to the flanges, e.g. flange cleats or an end plate connection, will almost certainly provide some warping restraint.

The imprecise nature of this problem is reflected in the very approximate way in which it is treated in BS 449. This first makes the approximation $k_1 = k_2 = k$ in Eq. 3.42 leading to the following interpretation of the effective length $\ell = kL$: 'the length of a beam of similar section subjected to similar loading which would have the same elastic critical moment as the beam in question'. Both versions of the code then use the following values of k:

1. Ends unrestrained against lateral bending $k = 1·0$
2. Ends partially restrained against lateral bending $k = 0·85$
3. Ends practically fixed against lateral bending $k = 0·70$

There are two reasons for this apparently conservative choice: first, practical end supports are unlikely ever to be capable of providing complete fixity against rotation and warping and second, recognition of the effect of different types of loading (e.g. for a central point load applied at the level of the centroid the theoretical value of k for complete fixity is approximately 0·63).

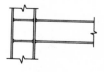

Fig. 3.32

Example 3.XIII

What would be the effective slenderness of the beam in Example 3.VII if both ends were completely restrained?

Using $0·70\lambda$ in Eq. 3.34 gives

$$\lambda_{LT} = \frac{0·859 \times 0·70 \times 129}{\sqrt[4]{\left[1 + \frac{1}{20}\left(0·70 \times 129\, \frac{1}{44·0}\right)^2\right]}}$$

$$= 74$$

The corresponding value of p_b is 166 N/mm² giving an increase in moment capacity over the simply supported case of 43%.

It is, of course, also possible that support arrangements will be encountered which provide less restraint than that shown in Fig. 3.29.

Studies of the amount of torsional restraint necessary at the end of the beam in order that it can safely be assumed that twisting is prevented suggest a figure of about twenty times the torsional stiffness of the beam itself. It is likely that this requirement would not normally be difficult to satisfy; in those doubtful cases, e.g. the eaves of a portal frame, methods exist for making a more detailed assessment.

The situation sometimes occurs (more frequently in bridge structures than in building structures) in which, as shown in Fig. 3.33, only the beam's lower flange is supported laterally. In such cases the restraint against lateral deflection of the top flange is provided only by the bending stiffness of the web and, especially for deep girders, this may be inadequate to prevent the sort of cross-sectional distortion shown in Fig. 3.33. Not surprisingly this phenomenon is associated with a reduced capacity. Two methods for dealing with this problem are included in the draft code. These are necessary because the problem occurs in what is essentially two different forms with an overlap or interactive region where elements of both are present. For slender beams the problem is one of the lateral-torsional buckling of the whole beam and the effect of the end conditions is simply that they do not provide the same measure of restraint as does the arrangement of Fig. 3.29. On the other hand for stocky beams the problem is essentially one of vertical buckling of the web as a strut in the region of the end reaction. Therefore the draft code requires the use of an effective

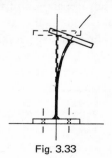

Fig. 3.33

length of $L + 2D$ when checking lateral buckling strength together with an additional check on the capacity of the web.

An alternative approach to this type of problem would be to provide vertical stiffeners at the support point thus preventing the type of distortion shown in Fig. 3.33.

Example 3.XIV

What would be the effective slenderness of the beam in Example 3.VII if it were provided with the sort of end support condition shown in Fig. 3.33, i.e. the top flange left unrestrained at the bearings?

effective length $\ell = L + 2D$

$$= 4000 + 899 \cdot 6 \text{ mm}$$

$$= 4900 \text{ mm}.$$

$$\therefore \quad \lambda = 4900/31 \cdot 1$$

$$= 158$$

and

$$\lambda_{LT} = \frac{0 \cdot 859 \times 158}{\sqrt[4]{\left[1 + \frac{1}{20} \left(158 \frac{1}{44 \cdot 0} \right)^2 \right]}}$$

$$= 120$$

leading to a design strength of 91 n/mm², a reduction of 22% over the simply supported case.

3.10 Behaviour of Continuous Beams

In certain circumstances it may be appropriate to treat a beam as being continuous over a number of spans. Depending upon the support arrangements continuity may be provided in the transverse plane, in the lateral plane or in both planes. Of these the first, i.e. a situation in which the reaction points do not also properly prevent lateral deflection and twisting, is both unusual and undesirable. Situations will sometimes occur in which this is unavoidable, e.g. roof purlins before sheeting or beams in temporary works. For such cases it is largely a matter of judgement as to whether any lateral support is being provided and in case of doubt the safest course is clearly to make no assumptions about possible restraint and therefore to design for the maximum effective length (including the possibility of end restraints that provide less than simple support, see Section 3.9).

Of rather more interest in connection with the problem of lateral buckling is the case of a single-span beam which is divided into several segments in the lateral plane by means of fully effective braces,* i.e. a beam that is continuous in the lateral plane. For such beams the buckled shape will involve deformation of all segments even though some of these may not be directly loaded. Provided each segment is of similar length and is subject to a similar pattern of moments in the transverse plane, then its effective length will be close to the spacing of the braces, but where the spacing is unequal or the moment pattern varies the effective length of each segment may be either more or less than its actual length.

As an example of this type of problem consider the single-span beam shown in Fig. 3.34. This is divided into three segments by the two equally spaced, equally loaded cross-beams and clearly both the critical moment of the main beam and the associated buckling mode will be dependent upon the spacing of these cross-beams. Moreover, this critical moment (M_{crB}) must, for any value of the ratio L_1/L_B, lie somewhere between the values of the buckling moments of the individual segments (M_{cr1} and M_{cr2}). Noting from Fig. 3.22 how for the outer segments the appropriate value for $1/m$ is $1\cdot75$, enables these to be obtained from Eqs. 3.25 and 3.40 as

$$M_{cr1} = 1\cdot75 \frac{\pi}{L_1} \sqrt{(EI_yGJ)} \left/ \left(1 + \frac{\pi^2 EI_w}{L_1^2 GJ}\right)\right.$$

$$M_{cr1} = \frac{\pi}{L_2} \sqrt{(EI_yGJ)} \left/ \left(1 + \frac{\pi^2 EI_w}{L_2^2 GJ}\right)\right. \tag{3.43}$$

A plot of M_{cr1} and M_{cr2} against L_1/L_B is given in Fig. 3.34. This shows how for an arrangement corresponding to a value of L_1/L_B of $0\cdot37$ M_{cr1} and M_{cr2} are equal, which means that all segments are simultaneously critical and the beam buckles with no interaction between adjacent segments. However, for any other value of L_1/L_B the more stable segment(s) will restrain the critical segment(s) leading to a value for the critical load which is greater than the lower of M_{cr1} or M_{cr2} as shown by the solid line in Fig. 3.34. Thus for values of L_1/L_B of less than $0\cdot37$ the outer segments will restrain the central segment, a situation that will be reversed when L_1/L_B exceeds $0\cdot37$. Viewed in terms of the effective length of the central segment ℓ_2, then as shown in the plan views of the buckled shapes given in Fig. 3.34, this will be less than its actual length L_2 when L_1/L_B is less than $0\cdot37$, it will be equal to L_2 at the 'zero interaction' value of L_1/L_B of $0\cdot37$, and it will exceed L_2 when the outer segments become critical for values of L_1/L_B of more than $0\cdot37$.

The results of Fig. 3.34 suggest that a safe value for the critical load of a laterally continuous beam may be obtained by taking the lowest value of the critical load of any individual segment calculated assuming that segment is simply supported at its ends. Research has shown this to be generally true, the exceptions being those arrangements where buckling occurs with no interaction in which case the process gives an 'exact' result.

*For a discussion of bracing requirements see Section 3.11.

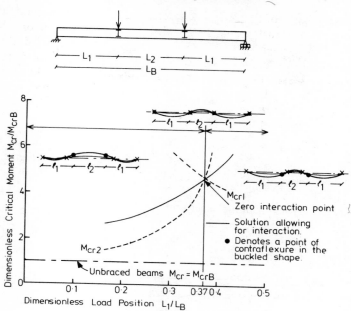

Fig. 3.34

Example 3.XV

A 457 × 152 UB 52 is simply supported over a span of 7·5 m as shown. It is laterally restrained at both ends as well as at both the load points. Estimate the critical value of P.

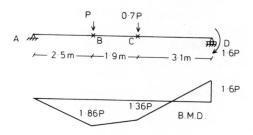

AB $\beta = 0{\cdot}0/1{\cdot}86\,P$ $= 0{\cdot}0$

  $m = 0{\cdot}57 + 0{\cdot}33(0) + 0{\cdot}1(0)^2 = 0{\cdot}57$

Using Eqs. 3.41 and 3.25, M = 871 kNm
  Corresponding value of P_{cr} = 469 kN

BC $\beta = 1{\cdot}37\,P/1{\cdot}86\,P$ $= 0{\cdot}737$

  $m = 0{\cdot}57 + 0{\cdot}33(0{\cdot}737) + 0{\cdot}1(0{\cdot}737)^2 = 0{\cdot}87$

Using Eqs. 3.41 and 3.25, M_{cr} = 957 kNm

Corresponding value of P_{cr} = 515 kN

CD $\beta = 1\cdot37\ P/(-1\cdot60\ P) = -0\cdot856$

  m = 0·43

Using Eqs. 3.41 and 3.25, M_{cr} = 780 kNm

Corresponding value of P_{cr} = 488 kN

Since the lowest value of P_{cr} is that for segment AB this segment will be the critical segment. Moreover, since the values of P_{cr} for the other two segments are not much higher the effects of interaction will be slight, segments BC and CD will not provide much restraint to segment AB, and the lower bound estimate will be only slightly less than the exact value.

The behaviour of beams which are continuous in both the transverse and lateral planes is essentially similar to that already described for laterally continuous beams, the only real difference being in the more complex patterns of

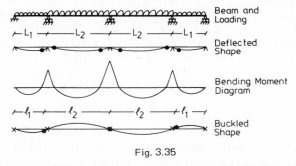

Fig. 3.35

moments that arise, see Fig. 3.35. This type of problem does, however, introduce the possibility of confusion over the buckled shape of the beam and in particular over the use of the term 'points of contraflexure'. It is most important to differentiate between points of contraflexure in the *buckled shape*, i.e. in the lateral plane, and points of contraflexure in the *deflected shape*, i.e. in the transverse plane. These will not normally occur at the same points within an individual span; indeed for the common example of a multispan beam with all spans uniformly loaded there will be considerably less of the former than of the latter, see Fig. 3.35. Thus it is quite wrong to use distances between points of zero moment as effective lengths for checking lateral buckling strength. In this respect the use of the term 'effective length of compression flange' often found in codes is misleading since it is, of course, the whole beam that is involved in the buckling.

3.11 Effective Lateral Restraint

The capacity of any beam whose design has been based upon considerations of lateral instability may be increased (subject, of course, to an upper limit of its in-plane strength) by the proper use of a system of lateral bracing. Such systems may consist either of discrete braces as in the case of a series of cross-beams or of a continuous lateral restraint as in the case of a beam partially embedded in a concrete floor. Assuming that the restraint system is fully effective, i.e. it is capable of acting as though it were completely rigid, then for the first type of arrangement the main beam will behave as a laterally continuous beam and its capacity can be checked using the methods of Section 3.10, whilst for the second class of problem the restraint system will normally be capable of completely preventing lateral buckling and design can be based upon consider-ations in in-plane behaviour alone. (An exception to this could occur for example in the hogging moment region of a continuous deep beam where restraint was provided to the tension flange and a form of lateral buckling involving cross-sectional distortion similar to that discussed in Section 3.9 could occur.)

3.11.1 Discrete Bracing

The behaviour of a beam provided with an arrangement of discrete braces may best be understood by means of a particular example. Consider the beam shown in Fig. 3.36 which is provided with a single, elastic lateral support of deflectional stiffness K_b. The relationship between the elastic critical moment M_{cr} of the beam and the value of K_b is shown in Fig. 3.37. Increasing K_b from zero produces a corresponding increase in M_{cr} up to the point at which K_b is equal to some limiting value K_{b1} and the corresponding value of M_{cr} becomes equal to the value for buckling in two half-waves. Further increases in K_b will not produce any further increase in M_{cr}, since the *lowest* buckling load is now that associated with a mode which does not involve deformation of the bracing.

Since the results of Fig. 3.37 were obtained from a bifurcation type of analysis, they yield no information on the magnitude of the buckling deflections and therefore no information on the forces induced in the bracing. This can only be obtained by basing the analysis on the behaviour of an initially bowed or twisted member. When this is done, it is found that, provided the beam is not designed to operate at loads close to its elastic critical load, then the forces induced in the bracing are not normally large. Furthermore, although the value of K_b required to produce a given increase in stability is now greater than was the case for the perfect beam, it is still of the same order of magnitude. This ability of comparatively light bracing to provide substantial increases in stability has also been confirmed by tests.

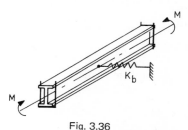

Fig. 3.36

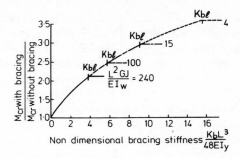

Fig. 3.37

Since the stiffness of lateral bracing necessary for it to act as if rigid is not normally particularly large, the most suitable and certainly the simplest basis for design consists of ensuring that all bracing members are at least as stiff as this. Clearly the number of possible arrangements of lateral bracing is extremely large. Taking as an example the case of a single-span beam provided with a single central brace it has been found that the value of $K_{b\ell}$ is affected by the following: the level of attachment of the brace to the beam, i.e. top flange, bottom flange, etc., whether the connection can resist moments in which case the brace will provide torsional as well as lateral restraint, the type of loading on the beam, especially the level of application of any transverse loads, the proportions of the beam, etc. It is possible, however, to identify undesirable situations and, provided these are avoided, to give simple rules for the proportioning of bracing members. Since such rules are expected to cover a range of situations they cannot be expected to be precise. However, this should not matter unduly, since they are not normally difficult to satisfy.

Arrangements in which transverse loads which are free to deflect sideways on buckling are applied above the level of attachment of the bracing, e.g. a crane trolley running on the top flange of a beam braced on the bottom flange, are particularly undesirable. Indeed tension flange bracing is generally not as effective as compression flange bracing, although methods do exist (for example for the design of columns in portal frames) which enable the stabilising influence of sheeting rails attached to the tension flange to be included. However, in many cases of beams, bracing attached below the level of application of transverse loads has been found to be incapable of raising the capacity of the beam to that of the fully braced case (because twisting of the braced cross-section is still able to occur).

From the foregoing discussion two essential design requirements may be identified for any bracing system:

(i) Sufficient stiffness to limit movement at the braced point(s), thereby ensuring that buckling occurs between the braces only.
(ii) Sufficient strength to withstand the forces transmitted to it by the main beam.

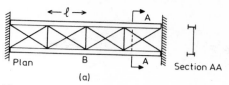

(a)

The plan bracing may be taken as fully effective and an effective length ℓ used providing the horizontal deflection at B due to a unit horizontal load is less than 1/25 of the deflection calculated assuming no bracing.

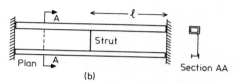

(b)

Providing Iy of the box beam is at least 25 times Iy of the I-beam then the effective ℓ for the latter is as shown.

Fig. 3.38

In the 1969 version of BS 449 only the second aspect was covered, the relevant clause requiring that the bracing system be capable of resisting a lateral force equal to 2½% of the maximum force in the compression flange of the main beam. This rule has been retained in the new draft code, where it has been supplemented by the addition of a bracing stiffness rule which requires that the stiffness of the bracing system be at least 25 times the lateral bending stiffness of the member to be braced before it can be considered as fully effective. Two examples of the use of this second rule are shown in Fig. 3.38. In making use of this rule it is, of course, necessary to consider exactly how the bracing is acting and what features contribute to its flexibility. Thus in Fig. 3.38(b), whilst the axial stiffness of the strut (which is the bracing directly connected to the beam being braced) may be expected to be very large, it is the lateral bending stiffness of the second beam that provides the major contribution to the flexibility of the system. Equally if the criss-cross part of the bracing shown in Fig. 3.38(a) had been omitted, the beneficial effect of the bracing would be greatly reduced, being dependent solely upon the ability of the three cross-members to inhibit twisting, which in turn would be a function of the bending stiffness of these members and their connections.

3.11.2 Continuous Restraint

When a beam is provided with fully effective, continuous lateral restraint then providing such restraint acts at or above the level of the beam's centroid* lateral buckling cannot occur. Thus for any simply supported beam whose top flange is

*It is assumed that the top flange is the compression flange.

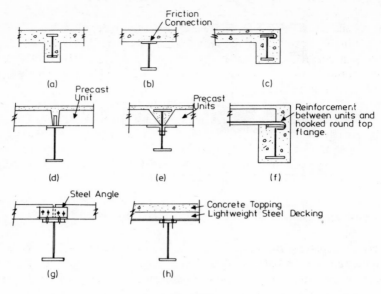

Fig. 3.39

effectively held in position laterally, design may be based upon considerations of in-plane strength alone. This situation will normally be found to exist in the majority of floor constructions and a series of typical details capable of providing full lateral restraint is shown in Fig. 3.39. In all of these cases it is, of course, necessary to appreciate that the restraint will be present only in the completed floor and that during construction temporary bracing may be required.

For the system shown in Fig. 3.39(a), in which the beam is fully encased in concrete, the concrete floor slab will normally be sufficiently rigid to prevent any movement of the compression flange. The arrangement of Fig. 3.39(b), in which the slab is cast directly on to an unpainted steel beam, will normally provide full restraint provided the load from the slab and the width of the beam flange are adequate to develop sufficient frictional resistance. A coefficient of friction of 0·3 may be used in determining the restraining force. For the case shown in Fig. 3.39(c) the bars hooked round the top flange should be capable of developing the full required restraining force, otherwise there is a danger of the concrete spalling in the region of the beam flange. Precast concrete floor units laid directly on to the beam flange will not, in themselves, normally provide the necessary resistance. However, if shear plates about 450 mm long spaced at 100 mm centres are used then when the whole system is grouted the resistance will usually be adequate. In the construction of Fig. 3.39(f) reinforcement placed between the units and hooked round the compression flange will, if designed to carry the full restraining force, be adequate. When the floor consists of timber joists as shown in Fig. 3.39(g), then the use of steel angles which are bolted through the timber beams and to the beam flange will normally prove satisfactory. Whilst the joists are usually placed at 400–600 mm centres, these can be considered as being sufficiently close together to be classed as continuous support. For the construction shown in Fig. 3.39(h) the thin metal decking, if used without concrete infill, may not possess sufficient shear stiffness to prevent the compression flange of the beam from moving laterally, particularly if it is only intermittently fastened.

Clearly it will not always be possible to use an arrangement similar to one of those shown in Fig. 3.39. Both the existing version of BS 449 and the new draft include a requirement that in order to be considered fully effective continuous lateral bracing must be capable of withstanding a force of 2½% of the maximum force in the compression flange of the beam taken as a uniformly distributed lateral load.

3.12 Cantilever Beams

Probably the most obvious example of a situation in which neither the support conditions nor the pattern of moments corresponds to the basic case is the cantilever beam. Research has shown that, with the exception of the rather unusual case of a cantilever carrying a moment at the tip, the most severe loading condition normally corresponds to a point load acting at the tip. Since the support conditions (in the transverse plane) directly affect the moment pattern it is convenient when attempting to provide design-oriented approaches for cantilevers to combine both loading and support effects and to use the idea of a 'notional effective length'.* This approach is used in both versions of the code, although the new draft code includes rather more cases. It has the advantage over an equivalent presentation in terms of m-factors that the values of k are generally less dependent on the value of $L^2 GJ/EI_w$. Fig. 3.40 gives recommended values of k for a number of cases.

At first sight the use of a value of k of less than 1·0 for the basic cantilever may seem strange but it should be remembered that the 'notional effective length' makes allowance for both the condition of support and the pattern of moments in this case. For simple cantilevers it has been found that prevention of twist at the tip is more effective than prevention of lateral deflection and this is reflected in the k-values quoted.

Cantilever spans will often occur as the overhanging span of an otherwise continuous (in the transverse plane) beam. In such cases the conditions of lateral restraint at the root have been found to be of less significance than the conditions of lateral restraint at the fulcrum (the most outward point of vertical support). Of particular importance is the effective prevention of twist at the fulcrum; failure to ensure this causes substantial reductions in lateral stability as shown by the greatly increased k-values.

*The notional effective length is defined as the length of the notional simply supported (in the lateral plane) beam of similar section, which would have an elastic critical moment under uniform moment equal to the elastic critical moment of the actual beam under the actual loading conditions.

Restraint Conditions		Loading Condition	
At Support	At Tip	Normal	Destabilising
Built-in laterally and torsionally	Free	0.8L	1.4L
	Lateral restraint only	0.7L	1.4L
	Torsional restraint only	0.6L	0.6L
	Lateral and torsional restraint	0.5L	0.5L
Continuous, with lateral and torsional restraint (see notes)	Free	1.0L	2.5L
	Lateral restraint only	0.9L	2.5L
	Torsional restraint only	0.8L	1.5L
	Lateral and torsional restraint	0.7L	1.2L
Continuous, with lateral restraint only (see notes)	Free	3.0L	7.5L
	Lateral restraint only	2.7L	7.5L
	Torsional restraint only	2.4L	4.5L
	Lateral and torsional restraint	2.1L	3.6L

Note: For continuous cantilevers $\ngtr L_1$, where L_1 is the length of the adjacent span.

Fig. 3.40

Example 3.XVI

Determine the value of λ'_{LT} for a 457 × 152UB52 beam when it is supported as shown below assuming:

(i) point B is fully braced and (ii) point B is only restrained against horizontal movement. The load is applied at C by means of a cross-beam framing in at the level of the neutral axis in such a way that it may be assumed to be capable of preventing twisting only.

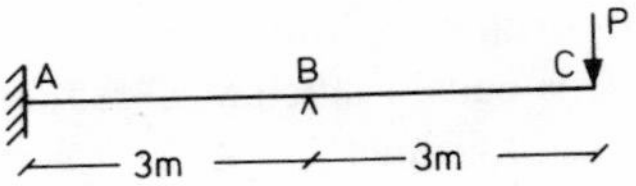

(i) From Fig. 3.40 effective length factor for BC is 0·8. However, since both spans are of equal length, compliance with the footnote requires the use of an effective length of not less than the full span AB (3 m in this case). Thus

$$\lambda = 300/3·11 = 96$$

From Eq. 3.34

$$\lambda'_{LT} = \frac{0.859 \times 96}{\sqrt[4]{\left[1 + \frac{1}{20}\left(96 \times \frac{1}{44.0}\right)^2\right]}} = 78$$

(ii) From Fig. 3.40 effective length factor for BC = 2.4.

$$\therefore \quad \lambda'_{LT} = \frac{0.859 \times 2.4 \times 96}{\sqrt[4]{\left[1 + \frac{1}{20}\left(2.4 \times 96 \times \frac{1}{44.0}\right)^2\right]}} = 159$$

When the loading is applied to the top flange of a cantilever then, particularly for beams possessing relatively low values of $L^2 GJ/EI_w$, very large reductions in lateral stability are produced. Greatly increased notional effective length factors are required in such cases, see Fig. 3.40. The values quoted for the case of destabilising loads and prevention of twist at the tip relate to uniformly loaded cantilevers since altering the level of application of a point load would have no effect in such cases. Data on this aspect of the problem were not available at the time of the 1969 version of the code was first written and specific guidance is provided only in the new draft code.

3.13 Local Buckling Effects

Apart from the brief reference to the possible occurrence of web distortion in cases where lateral support was provided only to the bottom flange, see Section 3.9, it has been assumed throughout this chapter that the cross-section maintains its original shape under load. However, since the types of members being considered are such that they consist of a number of relatively thin plates it is necessary to check that the proportions of these plates will not be such that local buckling effects will reduce the strength of the member below that calculated on the basis of overall bending behaviour.

The most convenient method of dealing with local buckling problems in design is quite simply to ensure that the members are so proportioned that local buckling effects cause no appreciable reduction in strength. For an I-section beam this requires that the width to thickness (b/t) ratios of all component plates, i.e. the web and the flanges, should not exceed certain limits, the exact values of which will depend upon whichever of the following three performance requirements is being considered:

(i)   Elements in members designed to just reach yield.
(ii) Elements in members designed to reach M_p, i.e. a beam with an effective lateral-torsional slenderness of less than λ_{LTo} (see Fig. 3.20).
(iii) Elements in members designed to participate in plastic hinge action.

The essential difference between each of these cases is the order of strain that the element in question must be capable of accepting before it buckles. Whereas in the first case the most heavily strained fibres need only reach ϵ_Y, for both the other cases strains of several times ϵ_Y must be developed.

Thus in the case of flanges designed on an elastic basis the new draft code uses the limit

$$B/2 \not> 16T\sqrt{(240/p_Y)} \qquad (3.44)$$

with the 16 being reduced to 14 for welded sections in recognition of the reduced local buckling strength of such sections.

Although cases (ii) and (iii) cover somewhat different situations (in case (ii) the member must just reach M_p without the appearance of significant local buckles, whilst in case (iii) it must possess a sufficiently flat-topped moment versus in-plane rotation curve for it to be suitable for plastic design), research has shown the same limitation on outstand dimensions to be suitable for both cases and the new code uses

$$B/2 \not> 10T\sqrt{(240/p_Y)} \qquad (3.45)$$

with the 10 being replaced by 9 for welded sections.

All UB and UC sections meet the requirements of Eq. 3.44 for all grades of steel; consequently Eq. 3.44 need only be checked when fabricated sections are employed. However, many rolled sections do not meet the far stricter limits of Eq. 3.45, and clearly such sections should not be used for plastic design. When used in a situation corresponding to (ii) above, however, the new code permits them to be designed using a maximum moment capacity obtained by linear interpolation between the values $p_Y S_x$ and $p_Y Z_x$ as shown in Fig. 3.41. (Note that this Figure is drawn for a particular value of shape factor $s = S_x/Z_x$.) This process replaces the earlier type of code requirements in which members that did not satisfy the stricter limit could only be designed for the capacity corresponding to the more lenient limit, i.e. a sudden reduction was employed rather than the pro rata reduction as shown in Fig. 3.41.

Similar checks are also necessary for the web. Depending on both the exact nature of the applied loading as well as the position within the beam's span the web may be subjected to various combinations of bending and shear. Research

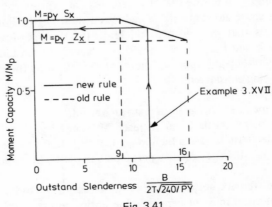

Fig. 3.41

on plate buckling has shown that the condition of pure shear is the most severe and that a web which meets slenderness requirements based on this case may be expected to perform satisfactorily in all situations. Once again the new draft code recognises different performance requirements, using

$$d \not> 135t \ \sqrt{(240/p_Y)} \ \text{for case (i)}$$

and

$$d \not> 110t \ \sqrt{(240/p_Y)} \ \text{for cases (ii) and (iii)} \qquad (3.46)$$

For depths greater than those given by Eq. 3.46 it is necessary either to accept a reduced strength or to employ web stiffeners.

Example 3.XVII

What is the moment capacity of the built-up I-section shown below when it is used as a beam with lateral bracings at $1 \cdot 28$ m centres? Use $p_Y = 340 \ N/mm^2$.

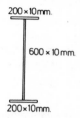

200×10mm.

600 × 10 mm.

200×10mm.

From Eq. 3.36 u = 1·0 and x = 620/10 = 31·0

$$\therefore \quad \lambda = 128/3 \cdot 65 = 35$$

and

$$\lambda_{LT} = \cfrac{1 \cdot 0 \times 35}{\sqrt[4]{\left[1 + \cfrac{1}{20} \left(35 \times \cfrac{1}{31 \cdot 0} \right)^2 \right]}} = 34 \cdot 5$$

From Table 6.3.2 of Draft Code for $p_Y = 340 \ N/mm^2$ corresponding value of $p_b = 320 \ N/mm^2$.

$\therefore$ based on considerations of lateral stability

$$M_b = 320 \times 2120 \ \text{Nm}$$

$$= 678 \ 400 \ \text{Nm}$$

Since this is quite close to $M_p = 340 \times 2120 = 720\ 800$ Nm it is necessary to check local flange stability against Eq. 3.46.

From Eq. 3.45 limiting value of B/2 = 9 x 10 $\sqrt{(240/340)}$

$$= 76 \ \text{mm}$$

Actual value of B/2 = 100 mm

∴ moment capacity based on local buckling strength of the flange needs to be determined using Fig. 3.41.

From Eq. 3.44 maximum value of B/2 for which M_Y may be obtained

$$= 14 \times 10 \sqrt{(240/340)} = 118 \text{ mm}$$

∴ for B/2 = 100 moment capacity $= M_Y + (M_P - M_Y) \dfrac{118 - 100}{118 - 76}$

$$= 340(1781 + (2120 - 1781) \times 0\cdot429)$$

$$= 340 \times 1927$$

$$= 665\,000 \text{ Nm}$$

This calculation is also displayed on Fig. 3.41 (B/2T $\sqrt{(240/p_Y)} = 11\cdot5$ giving $M/M_p = 0\cdot924$).

From Eq. 3.47 limiting value of $d = 110 \times 10 \sqrt{(240/340)}$

$$= 924 \text{ mm}$$

Actual value of d = 600 mm

∴ web is satisfactory.

Thus in this example the beam's capacity is limited by the local buckling strength of its flanges to 665 000 Nm.

3.14 Beams other than Uniform, Equal-Flanged I-Sections

The theory of the twisting and buckling of I-beams developed in Sections 3.2 and 3.3 was based on the assumption of a doubly symmetrical cross-section having constant properties throughout its length. Although many of the lateral buckling problems commonly encountered in design fall within this category, situations will arise in which the stability of other types of beam needs to be assessed. Since the treatment of sections which are not both uniform and doubly symmetrical involves modifications to the basic theory it is not possible in this introductory text to cover such problems in detail. Rather some qualitative indications of the behaviour of such members will be presented as a conclusion to this chapter.

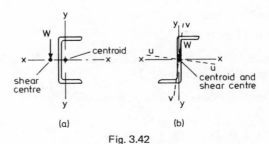

Fig. 3.42

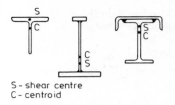

S - shear centre
C - centroid

Fig. 3.43

Symmetrical channels and zeds will only exhibit lateral instability if the applied loading acts in such a way as to produce pure major-axis bending. For the former this requires the loads to act parallel to the web through the shear centre as shown in Fig. 3.42(a). Since zeds possess point symmetry with the principal axes being inclined to the normal xx and yy axes, true lateral instability can only occur if the loading is applied as shown in Fig. 3.42(b). Provided these conditions are satisfied the value of M_{cr} may be obtained in either case directly from Eq. 3.25 by inserting the appropriate values of I_y, J and I_w where the value of I_w is given by

$$I_w = \frac{TB^3 h}{12} \left(\frac{3BT + 2ht}{6BT + ht} \right) \text{ for a channel}$$

$$I_w = \frac{B^3 h^2}{12(2Bth)^2} [2t(B^2 + Bh + h^2) + 3tBh] \text{ for a zed}$$

In practice the loading arrangements will often not conform to the restrictions given above in which case it becomes a matter of judgement as to whether the problem is one of lateral instability or combined biaxial bending and twisting.

For sections which are symmetrical about the minor axis only, such as those shown in Fig. 3.43, the shear centre does not lie on the xx-axis. This complicates the torsional behaviour and Eq. 3.12 can no longer be applied with the result that Eq. 3.25 is not valid for this type of section. Some quantitative indication of the effect on lateral stability of varying the relative proportions of the flanges may be obtained from Fig. 3.44, which shows clearly the beneficial effect of concentrating material in the compression flange. Procedures for the design of such beams make due allowance for the effects shown in Fig. 3.44.

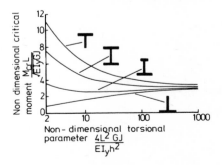

Fig. 3.44

Curtailment of a beam's cross-section in order to match the moment diagram is a device which is sometimes employed. Studies of the effects of taper on lateral stability show that whereas variations in flange proportions can produce substantial reductions in the buckling load, depth taper (which is, of course, a convenient and popular means of obtaining a varying major-axis section modulus) has relatively little effect. This observation forms the basis for the simplified methods for checking the lateral stability of non-uniform members given in design codes.

3.15 Bibliography

1. Timoshenko, S. P., and Gere, J. M. (1961), *Theory of Elastic Stability*, 2nd edn., New York, McGraw-Hill, ch. 6.
2. Bleich, F. (1952), *Buckling Strength of Metal Structures*, New York, McGraw-Hill, ch. 4.
3. Galambos, T. V. (1968), *Structural Members and Frames*, Englewood Cliffs, N.J., Prentice-Hall, ch. 3.
4. Trahair, N. S. (1977), *The Behaviour and Design of Steel Structures*, London, Chapman and Hall, ch. 6.
5. Chen, W. F., and Atsuta, T. (1978), *Theory of Beam-Columns*, vol. 2, New York, McGraw-Hill, ch. 3.
6. European Convention for Constructional Steelwork (1976), *Second International Colloquium on Stability – Introductory Report*, ECCS, ch. 5.
7. Johnston, B. G. (ed.) (1977), *Guide to Design Criteria for Metal Compression Members*, 3rd ed., SSRC, New York, Wiley, ch. 6.
8. British Standard 449, Part 2: 1969 (1969), *Specification for the use of Structural Steel in Building*, London, BSI.
9. British Standard 153, Parts 3B and 4: 1972 (1972), *Specification for Steel Girder Bridges*, London, BSI.
10. *Draft Standard Specification for the Structural Use of Steelwork in Building. Part I: Simple Construction and Continuous Construction* (1977), London, BSI.

CHAPTER FOUR

The Elastic Critical Loads of Plane Frames

4.1 Introduction

As pointed out in Chapter 2 most member stability problems really involve frame behaviour. When continuity of any form exists between the member under consideration and adjacent members then the determination of an elastic critical load (or of an effective length) which does not include a quantitative assessment of the interaction effects is an approximation.

In this chapter frames are considered under two basic headings:

(i) No-sway frames; where it is assumed that the ends of members are not free to move relative to each other, e.g. triangulated frames or multistorey portal frameworks with sway bracing.

(ii) Sway frames; where the resistance to lateral loads is provided by sway moments induced in the columns.

The stiffness of the joints themselves is an important parameter as it influences the degree of interaction which takes place. This text will be largely confined to in-plane buckling of frames with the two extreme cases of pinned or fully rigid joints.

4.2 Frame Behaviour and Frame Instability

Consider any statically determinate plane frame in which all joints are perfectly pinned, i.e. there is no rotational restraint transmitted from any one member to another. Suppose that the frame is loaded at the joints, then, as the loads are increased, instability will occur when the axial compressive force in any one member reaches the Euler load of that member, e.g. the frame of Fig. 4.1.

It would however be a rare occurrence for such a frame to be constructed. Joints in frames of this type are normally well able to transmit considerable moments and a more realistic approximation to the behaviour may be obtained by assuming complete rotational continuity at the joints, i.e. the ends of all members meeting at a joint rotate equally.

Consider the two-bar frame of Fig. 4.2 which is rigidly jointed at the apex, B,

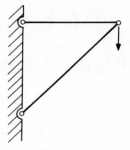

Fig. 4.1

and with joints A and C encastré. The axial forces in the members AB and BC may be estimated by assuming the joints to be pinned. The error involved in this assumption is slight whereas the simplification of the ensuing computation is great. These axial forces can be shown to be

(i) Force in AB, $P_{AB} = W \left[\dfrac{v}{L_1} + \dfrac{h_1 v}{L_1 h_2} \right]$

and

(ii) Force in BC, $P_{BC} = W \left[\dfrac{v}{L_2} + \dfrac{h_2 v}{L_2 h_1} \right]$

Thus the ratio of the axial loads may be established. If the member lengths (L_1 and L_2) and their in-plane flexural rigidities (EI_{AB} and EI_{BC}) are known then their Euler loads (P_{EAB} and P_{EBC}) and their relative stiffnesses (k_{AB} and k_{BC}) may also be determined.

The problem of calculating the in-plane elastic critical load is now very simple. A convenient test distortion to apply is a rotation of joint B, θ_B. Application of the modified slope deflection equation (Eq. 2.38) to both members meeting at the joint leads to the pattern of moments of Fig. 4.3. As joints A and C are fixed these are the moments which will occur in the frame due to the rotation of B. The only moments which are required to determine the resistance of joint B to the rotation are M_{BA} and M_{BC}. A measure of the resistance of the frame to the distortion is $\Sigma_B M$ which is M_{BA} plus M_{BC}.

$$\Sigma_B M = (k_{AB} s_{AB} + k_{BC} s_{BC}) \theta_B$$

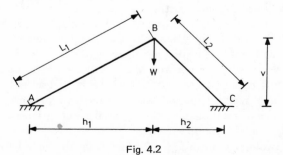

Fig. 4.2

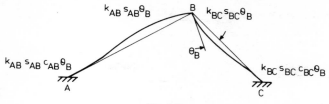

Fig. 4.3

When this quantity is positive the frame is in stable equilibrium, when it is negative the frame is unstable and when $\Sigma_B M$ is zero the loading is critical. The condition for instability is that

$$\Sigma_B M = k_{AB} s_{AB} + k_{BC} s_{BC} = 0 \qquad (4.1)$$

and the solution must be obtained by a trial and error process.

Example 4.1

Consider the frame of Fig. 4.2 with $L_1 = 3\cdot00$ m, $L_2 = 2\cdot12$ m, $h_1 = 2\cdot60$ m, $h_2 = 1\cdot50$ m, $I_{AB} = 1200$ cm^4 and $I_{BC} = 1000$ cm^4. ($v = 1\cdot50$ m)

Step 1 Calculate the forces in the members assuming pin-joints

$$P_{AB} = W \left[\frac{1\cdot50}{3\cdot00} + \frac{2\cdot60 \times 1\cdot50}{3\cdot00 \times 1\cdot50} \right] = W/1\cdot367$$

$$P_{BC} = W \left[\frac{1\cdot50}{2\cdot12} + \frac{1\cdot50 \times 1\cdot50}{2\cdot12 \times 2\cdot60} \right] = W/1\cdot116$$

Step 2 Determine the Euler loads and the relative stiffnesses

$$P_{EAB} = \frac{\pi^2 \times 207 \times 1200 \times 10^4}{3^2 \times 10^6} = 2725 \text{kN}$$

$$P_{EBC} = \frac{\pi^2 \times 207 \times 1000 \times 10^4}{2\cdot12^2 \times 10^6} = 4547 \text{kN}$$

$$\rho_{AB} = P_{AB}/P_{EAB} = \frac{W}{1\cdot367} \times \frac{1}{2725} = \frac{W}{3725}$$

$$\rho_{BC} = P_{BC}/P_{EBC} = \frac{W}{1\cdot116} \times \frac{1}{4547} = \frac{W}{5074}$$

Thus

$$\rho_{AB}/\rho_{BC} = 1\cdot36$$

$$k_{AB} = I_{AB}/L_1 = 1200 \times 10^4/(3 \times 10^3) = 4000 \text{ mm}^3$$

$$k_{BC} = I_{BC}/L_2 = 1000 \times 10^4/(2\cdot12 \times 10^3) = 4717 \text{ mm}^3$$

Thus

$$k_{AB}/k_{BC} = 4000/4717 = 0\cdot848$$

Step 3   Solve the instability criterion $\Sigma_B M = 0$ (Eq. 4.1)
$$0.848 s_{AB} + s_{BC} = 0$$

	Trial 1	Trial 2	Trial 3	Trial 4	Trial 5	Trial 6
ρ_{BC}	0	1·00	2·00	1·50	1·80	1·72
ρ_{AB}	0	1·36	2·72	2·04	2·45	2·34
s_{BC}	4	2·47	0·14	1·46	0·72	0·93
s_{AB}	4	1·76	−2·93	0·02	−1·52	−1·95
$0.848 s_{AB}$	3·39	1·49	−2·48	0·02	−1·29	−0·89
$0.848 s_{AB} + s_{BC}$	7·39	3·96	−2·34	1·48	−0·57	0·04

Step 4   Evaluate W_{cr}.

From the table it can be seen that when $\rho = 0$ joint B has a positive resistance to a rotation (proportional to 7·39 units). As ρ is increased this resistance falls, becoming zero (or very nearly so) when $\rho_{BC} = 1·72$ and $\rho_{AB} = 2·34$. Thus the frame becomes unstable when $\rho_{BC} = 1·72$, i.e. when $W_{cr} = 1·72 \times 5074 = 8730$ kN

or alternatively when $\rho = 2·34$, i.e. $W_{cr} = 2·34 \times 3725 = 8720$ kN.

The small discrepancy is due to round off errors. A crude approximation to this value could have been obtained as follows: having determined the axial forces in AB and BC suppose that the two bars have no moment interaction at B; then each would act as a strut pinned at one end and fixed at the other with critical loads of $2·05 P_E$ for each respectively. The value of the axial load in AB at which AB would become unstable is given by

$$P_{AB} = 2·05 \times 2725 = 5590 \text{ kN}$$

and the value of the axial load in BC at which BC would become unstable is given by

$$P_{BC} = 2·05 \times 4547 = 9320 \text{ kN}$$

Clearly BC is far more stable than AB and the former will tend to support the latter due to the rigid connection. Therefore

$$W_{cr}/1·367 > 5590 \text{ kN}$$

or

$$W_{cr} > 7640 \text{ kN}$$

Similarly AB will tend to weaken BC as it requires support.

Therefore

$$W_{cr} / 1·116 < 9320 \text{ kN}$$

or

$$W_{cr} < 10\,400 \text{ kN}$$

Combining these two conditions:

$$7640 \text{ kN} < W_{cr} < 10\ 400 \text{ kN}$$

Taking an average value $W_{cr} \simeq 9020$ kN.

This is a simple example of intuitive upper and lower bound techniques. In this case the bounds are quite widely spaced, differing by a factor of $1 \cdot 4$ but sometimes the bounds could be so close as to render further computation unwarranted. Even if the bounds are significantly different the process does enable first trial values of ρ to be selected on a rational basis.

Perhaps of more significance is the fact that the resulting elastic critical load is significantly different from those which might be obtained using Clause 31(a) of BS 449. Interpreting the construction to give effective restraint in position and direction at both ends the effective length would be taken as $0 \cdot 7L$. As the weaker member AB controls the design this would give rise to a critical force in AB given by

$$P_{AB} = \frac{\pi^2 \times 207 \times 1200 \times 10^4}{(0 \cdot 7 \times 3)^2 \times 10^6} = 5560 \text{kN}$$

This force in AB will occur when $W = 1 \cdot 367 \times 5560 = 7600$ kN which significantly underestimates the critical load even with a somewhat optimistic assessment of the restraint at B.

If a more conservative assessment regarding the restraint at B is adopted, i.e. assuming that AB is not effectively restrained in direction, then the effective length becomes $0 \cdot 85L$ which corresponds to a value of W_{cr} given by

$$W_{cr} = \frac{1 \cdot 367 \times \pi^2 \times 207 \times 1200 \times 10^4}{(0 \cdot 85 \times 3)^2 \times 10^6} = 5150 \text{kN}$$

The results may be summarised as follows:

True critical load	8720kN
Using an effective length of AB = $0 \cdot 7L$	7600kN
Using an effective length of AB = $0 \cdot 85L$	5150kN

The main purpose of this example, however, is to illustrate the difference between the action of the frame with and without rotational continuity, i.e. pinned or rigid joints. With all three joints pinned the system would become unstable when the load in one member reaches its Euler load, i.e. $\rho = 1 \cdot 00$. This would first occur in member AB when $W = 3725$ kN. With rotational continuity the critical condition corresponds to a value of $W = 8730$ kN.

It can be seen that rigidly jointed frames can have elastic critical loads very much higher than similar pin-jointed frames. The difference between the two values is due to the restraint offered by the members which frame into the most heavily loaded compression member. The Clauses of BS 449:1969 relevant to effective lengths offer a simple but crude (and frequently very conservative) method of estimating the elastic critical load of a framework. When checking the adequacy of an individual member of a framework the effective length may be determined from the value of ρ_{cr} for that member as set out below:

$$\rho_{cr} = P_{cr}/P_E = \frac{\pi^2 EI}{(\ell_{eff})^2} \bigg/ \frac{\pi^2 EI}{L^2} = \left(\frac{L}{\ell_{eff}}\right)^2$$

Thus

$$\ell_{eff} = \frac{L}{\sqrt{\rho_{cr}}} \tag{4.2}$$

There are ways in which elastic critical loads (and hence values of ρ_{cr}) may be determined without undue onerous computations and this chapter sets out to explain some of these techniques.

4.3 Elastic Critical Loads of No-Sway Frameworks: Patterns of Moments Technique

The basic philosophy previously described can be applied to deal with other triangulated plane frameworks. For example if the frame of Fig. 4.2 had joints A and C pinned instead of fixed it would have been necessary to consider patterns of distortion corresponding to rotations of A and C as shown in Fig. 4.4(a) and (b) respectively. These patterns must be combined with the pattern of Fig. 4.3 to eliminate the moments A and C as it is impossible for these joints to sustain any moment. Thus it is necessary to employ $-c_{AB}\theta_B/\theta_A$ times the pattern of Fig. 4.4(a) to eliminate the out-of-balance moment at A and $-c_{BC}\theta_B/\theta_C$ times the pattern of Fig. 4.4(b) to eliminate the out-of-balance moment at C which appears in Fig. 4.3. The total moment at B which then resists the rotation θ_B is given by

$$\Sigma_B M = [k_{AB}s_{AB}(1 - c_{AB}^2) + k_{BC}s_{BC}(1 - c_{BC}^2)]\theta_B$$

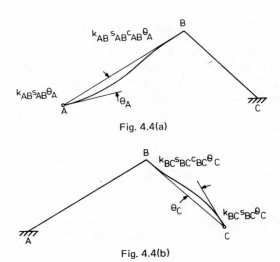

Fig. 4.4(a)

Fig. 4.4(b)

Thus for a member which is pinned at the remote end an effective stiffness equal to $s(1 - c^2)$ or s'' can be used in place of s with zero moment at the remote end. This is a direct equivalent to the ¾k used in normal moment distribution analyses. (Note that when $\rho = 0$, $s = 4$, $c = ½$ and $s(1 - c^2) = 4(1 - ½) = 4(¾)$ instead of s = 4.)

The stability criterion $\Sigma_B M = 0$ now becomes

$$\Sigma_B M = \frac{k_{AB}}{k_{BC}} \times s''_{AB} + s''_{BC} = 0 \qquad (4.3)$$

Compare this with Eq. 4.1.

Example 4.II

Consider the frame of Example 4.I but with joints A and C pinned against rotation.

The stability criterion now becomes

$$0.848 \, s''_{AB} + s''_{BC} = 0$$

which must be solved by trial and error as before.

	trial 1	trial 2	trial 3	trial 4	trial 5
ρ_{BC}	0	0·80	0·90	0·84	0·86
ρ_{AB}	0	1·09	1·22	1·14	1·17
s''_{BC}	3	0·86	0·46	0·71	0·63
s''_{AB}	3	−0·48	−1·31	−0·77	−0·97
$0.848 s''_{AB}$	2·54	−0·41	−1·11	−0·65	−0·82
$0.848 s''_{AB} + s''_{BC}$	5·54	0·45	−0·65	0·06	−0·19

Thus $\rho_{BC \, cr} \simeq 0.84$.

and thus

$$W_{cr} = 0.84 \times 5074 = 4260 \text{ kN}.$$

Note the fall in W_{cr} from Example 4.I

More extensive frameworks can be solved by using an extension of this method but the algebra involved rapidly becomes more complex. Nevertheless, a simple triangular frame will be solved using this approach.

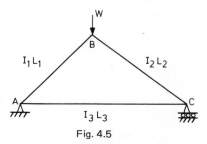

Fig. 4.5

In the triangulated framework of Fig. 4.5 all properties associated with members AB, BC and CA are given the suffixes 1, 2 and 3 respectively. The assessment of the axial load in each member (P_1, P_2 and P_3) is quite simple as is the determination of the Euler loads (P_{E1}, P_{E2} and P_{E3}). Hence the value of the ratio P/P_E may be found for all members (ρ_1, ρ_2 and ρ_3) together with k values (k_1, k_2 and k_3).

All possible distortions of the frame may be synthesised from the three basic distortion patterns:

(i)   Rotation of A through θ_A with θ_B and θ_C zero
(ii) Rotation of B through θ_B with θ_C and θ_A zero
(iii) Rotation of C through θ_C with θ_A and θ_B zero

The moment patterns corresponding to these basic disturbances are shown in Fig. 4.6(a), (b) and (c).

Suppose that joint B is chosen as the joint at which the test rotation is to be applied. (This choice in no way affects the expression which will subsequently be derived.) The pattern of Fig. 4.6(a) would be the correct pattern if, in the real frame, the ends A and C of the members BA and BC respectively were encastré. As this condition is not satisfied rotations will occur at both joints until no net

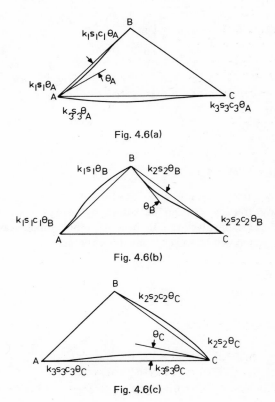

Fig. 4.6(a)

Fig. 4.6(b)

Fig. 4.6(c)

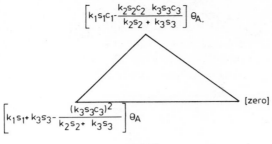

Fig. 4.6(d)

out-of-balance moment is present at either A or C. This is accomplished by combining the three basic patterns of Fig. 4.6(a), (b) and (c) in the correct ratios.

The first step is to combine patterns (a) and (c) in such proportions that the net moment at C is zero. This is done in the pattern of Fig. 4.6(d). This new pattern will subsequently enable a moment at A to be balanced without upsetting the equilibrium at joint C. Similarly the pattern of Fig. 4.6(e) in which the net moment at A is zero is obtained. The use of this pattern of moments will enable a moment at C to be balanced without disturbing equilibrium at A.

Patterns (d) and (e) are now employed (simultaneously or successively) on pattern (b) to eliminate the out-of-balance moments which occur at A and C due to the rotation of B with A and C clamped. The resulting pattern, shown in Fig. 4.6(f), is the moment pattern which will occur in the frame due to the rotation of B through θ_B with A and C free to rotate until equilibrium is obtained. The condition for instability is that the net moment at B is zero,

i.e.

$$(k_1 s_1 + k_2 s_2) - \frac{k_1 s_1 c_1}{\bar{S}_B} \bar{S}_{A1} - \frac{k_2 s_2 c_2}{\bar{S}_c} \bar{S}_{A2} = 0 \qquad (4.4)$$

where

$$\bar{S}_{A1} = k_1 s_1 c_1 - \frac{k_2 s_2 c_2 k_3 s_3 c_3}{k_2 s_2 + k_3 s_3}$$

$$\bar{S}_{A2} = k_2 s_2 c_2 - \frac{k_3 s_3 c_3 k_1 s_1 c_1}{k_3 s_3 + k_1 s_1}$$

$$\bar{S}_B = k_1 s_1 + k_3 s_3 - \frac{(k_3 s_3 c_3)^2}{k_2 s_2 + k_3 s_3}$$

$$\bar{S}_C = k_2 s_2 + k_3 s_3 - \frac{(k_3 s_3 c_3)^2}{k_3 s_3 + k_1 s_1}$$

This is an extremely cumbersome expression which must now be solved by trial and error. It is equivalent to finding the value of the load parameter at which the determinant of the stiffness matrix of the frame vanishes. This matrix

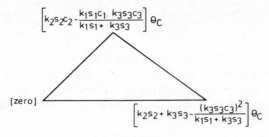

Fig. 4.6(e)

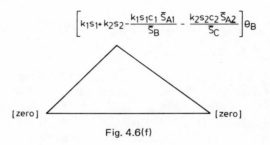

Fig. 4.6(f)

can be assembled from the first three moment patterns of **Fig. 4.6** as

$$
\begin{pmatrix} M_A \\ M_B \\ M_C \end{pmatrix} =
\begin{pmatrix}
k_3 s_3 + k_1 s_1 & k_1 s_1 c_1 & k_3 s_3 c_3 \\
k_1 s_1 c_1 & k_1 s_1 + k_2 s_2 & k_2 s_2 c_2 \\
k_3 s_3 c_3 & k_2 s_2 c_2 & k_2 s_2 + k_3 s_3
\end{pmatrix}
\begin{pmatrix} \theta_A \\ \theta_B \\ \theta_C \end{pmatrix}
$$

If the determinant of these coefficients was expanded it would be found to be identical with Eq. 4.4. It would be simpler however to determine the numerical values of the coefficients of the stiffness matrix for trial values of the load parameter and evaluate their determinant numerically. Alternatively if a desk-top computer were available the job may be easily programmed. The critical load occurs when the matrix first becomes singular, i.e. when the determinant of its coefficients first vanishes.

Fortunately it is not normally necessary for such a lengthy computation to be undertaken. Two of the alternative manual approaches which can be used on even more extensive frames such as that of **Fig. 4.7** will be described.

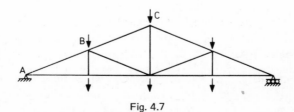

Fig. 4.7

4.4 Elastic Critical Loads of No-Sway Frameworks: Winter's Relaxation Technique

This method is due to Winter et al. and is a relaxation technique employed to evaluate the stiffness of a frame for several arbitrarily selected trial values of the load parameter. Initially the axial loads are calculated in terms of a general load parameter assuming the frame to be pin-jointed, the ratio of the axial load to the Euler load of each member is found and the information tabulated. Some particular trial value of this load parameter is selected thus enabling ρ, s and c to be determined for all members. From this information and the values of k for each member, the stiffness of each joint may be evaluated as the summation of the ks values for all members meeting at that joint. A suitable disturbance is applied to the frame which may for example be the rotation of a single joint — say the rotation of joint B. Having selected the test rotation the moments which arise on each member framing into the joint may be written down assuming that the remote ends of these members are fixed, i.e. these moments are proportional to the ks values of the members. Due to the test rotation, moments must also occur at the remote ends of the members framing into the tested joint and the magnitudes of these moments will be proportional to the value of ksc for that member. This is based on the assumption that the 'remote ends' are fixed (which is erroneous) and the moments arising here are out of balance. Permitting each of these 'unbalanced' joints to rotate will result in a rotation such that the total moment at the joint due to this balancing rotation is equal to —ksc and the proportion of this moment allotted to any one member again being proportional to the ratio of the ks value of the member concerned divided by the summation of the ks values of all members meeting at the joint.

The process is essentially one of normal moment distribution but with the modification that the stiffness of a member is dependent upon ks for that member (rather than its k value) and the carryover factors are evaluated as the appropriate value of c and not as one-half. Balancing and rebalancing of all joints except the one subjected to the test rotation is continued until sufficiently small residual out-of-balance moments are obtained (again compare with normal moment distribution). The moment which now remains at the tested joint is a measure of the stiffness of the frame with respect to the tested joint. The procedure is repeated for other values of the load parameter and a plot of stiffness versus load parameter obtained. When the load parameter is low the stiffness will be high. As the load increases the stiffness will fall, eventually becoming negative, and the value of the load parameter corresponding to zero stiffness gives the elastic critical load.

The choice of the disturbance to be applied is not normally of great significance provided it contains at least some component of the critical mode. Only with certain symmetrical frames is this likely to present any problem where certain joints on an axis of symmetry may not rotate at all in the lowest critical mode. (This could occur with joint C in the frame of Fig. 4.7.) Convergence is, however, usually more rapid if the 'weakest' joint of the frame is considered. For example if the two most heavily loaded compression members meet at a joint then (within the limitations of symmetry mentioned above) this joint is the most appropriate one for application of the test disturbance as it is likely that its stiffness will be the most sensitive to instability.

The mechanics of the method are best illustrated by use of an example.

Example 4.III

Consider the rigidity jointed framework of Fig. 4.8 in which all members are of the same cross-section with I = 333 cm⁴ and L = 4·00 metres. Find the value of W at which the frame would become elastically unstable in the absence of yielding.

Step 1   Estimate the axial loads in the members of the frame in terms of the load parameter W assuming all joints to be pinned. Column 1 of the table of Fig. 4.9.

Step 2   Determine the relative P_E values of all members. Column 2 of Table of Fig. 4.9. (As EI is constant throughout, the P_E values are inversely proportional to the square of the member lengths.)

Step 3   Evaluate the relative ρ values and the relative k values. Columns 3 and 4 of Table of Fig. 4.9.

Step 4   At this stage it is necessary to determine the joint to be tested. Examination of column 3 shows that DE is the most heavily compressed member. At E there are two compression members and at D there are three compression members and one tension member. There appears to be little to choose between the two alternatives and either is suitable. Select joint D.

Step 5   A trial value of loading must now be selected and a ρ value of 2 in member DE would appear convenient and somewhere in the vicinity of the critical load if one assumes that the end constraints of DE are such that it is roughly equivalent to a pin-fixed strut.

The s, ks and c values may be evaluated and are also entered into the table, as shown in columns 6, 7 and 8.

Distribution coefficients may now be evaluated. The coefficient $DC_{i \to j}$ is the moment which is transferred to joint j when unit balancing moment is applied at joint i. Remembering that the proportion of the unit moment arising on member

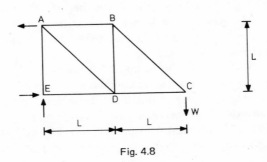

Fig. 4.8

Col. No	1	2	3	4	5	6	7	8
Member	Load	Rel P_E	Rel ρ	Rel k	Test ρ	s	ks	c
AB	−W	1	−1	1	−1·00	5·17	5·17	0·34
AD	−2W	0·5	−2√2	1/√2	−2·83	6·85	4·85	0·22
AE	W	1	1	1	1·00	2·47	2·47	1·00
BC	−W	0·5	−2√2	1/√2	−2·83	6·85	4·85	0·22
BD	W	1	1	1	1·00	2·47	2·47	1·00
CD	W	1	1	1	1·00	2·47	2·47	1·00
DE	2W	1	2	1	2·00	0·143	0·143	24·68

Fig. 4.9

ij at i is equal to $(ks)_{ij}/\Sigma_i ks$ and also that the carryover factor is c_{ij}, the distribution coefficient is given by

$$DC_{i \to j} = (ksc)_{ij}/\Sigma_i(ks) \tag{4.5}$$

where Σ_i indicates that the summation is carried out for all members meeting at i.

As an example the distribution coefficient from A to B is evaluated as follows:

$$(ksc)_{AB} = 5·17 \times 0·34 = 1·76$$

$$\Sigma_A(ks) = 5·17 + 4·85 + 2·47 = 12·49$$

$$DC_{A \to B} = 1·76/12·49 = 0·141.$$

All other distribution coefficients are computed in a similar manner. There are a total of fourteen distribution coefficients (two for each member), each being the moment carried over to the 'far' end of a member when unit balancing moment is applied to the joint at the 'near' end. These are summarised in the diagram of Fig. 4.10.

The test unit rotation at D is applied causing a total moment of Σks to appear at D and out-of-balance moments of ksc at the remote ends of all members meeting at D. Balancing of all joints except D is carried out using the distribution coefficients of Fig. 4.10 until satisfactorily small residuals are left. The Table of Fig. 4.11 illustrates this step. Similar calculations carried out at $\rho = 0$,

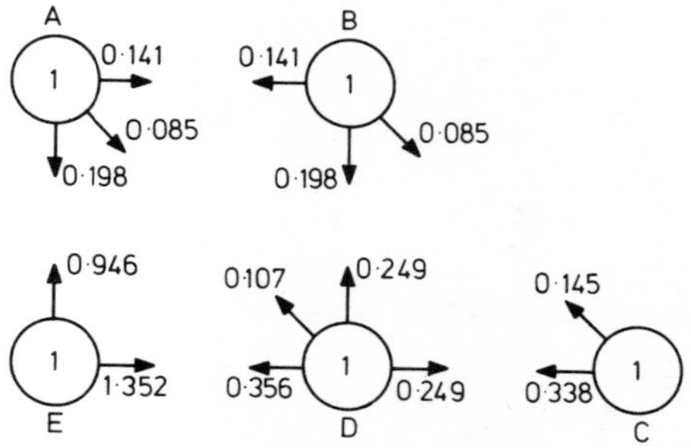

Fig. 4.10

Joint		A		B			C		D				E	
Member	AB	AD	AE	BA	BC	BD	CB	CD	DA	DB	DC	DE	EA	ED
Rot D.		1·06			−2·47	2·47		2·47	4·83	2·47	2·47	0·143		3·53
c.o.	−0·35		−3·34	−0·15			−0·21		−0·09			−4·77	−0·21	
Bal.	0·07	−1·06	0·20	0·52	−2·47		0·04	−2·47	0·31	−0·49	−0·83	0·28	0·73	−3·53
c.o.	−0·08	3·69	−0·69	−0·04	−0·36		−0·05	0·21	−0·02	0·10	0·07	−0·99	−0·05	0·21
Bal.	0·01	−0·27	0·05	0·11	0·51		−0·01	−0·04	0·07	−0·11	−0·01	0·07	0·15	−0·73
c.o.	−0·02	0·77	−0·14	−0·01	0·03			0·05	−0·01	0·01	0·02	−0·20	−0·01	0·05
Bal.		−0·06	0·01	+0·02	−0·55			0·01	0·01	−0·02		0·01	0·03	−0·15
c.o.		0·16	−0·03		−0·01							−0·04	0·01	0·01
Bal.		−0·01	−0·01		0·05									−0·03
c.o.		0·03			0·01									−0·01
Bal.		0·01			−0·12									
c.o.					0·01									
Bal.					−0·02									

$$\Sigma = 3·27$$

Fig. 4.11

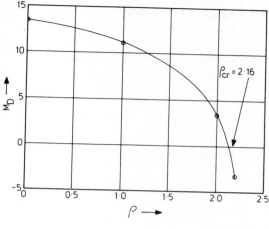

Fig. 4.12

$1\cdot0$, $2\cdot20$ give values of M_D of $13\cdot45$, $11\cdot05$ and $-3\cdot28$ which when plotted on to Fig. 4.12 give a critical value of ρ in member DE of $2\cdot16$.

Now the Euler load of DE is given by

$$P_{E\,DE} = \pi^2 \times 207 \times 10^{-2} \times 333/4^2 = 425 \text{ kN}$$

The axial load in DE is 2W and therefore

$$2W_{cr}/425 = 2\cdot16$$

or

$$W_{cr} = 459 \text{ kN}$$

The degree of precision obtainable by this lengthy approach is rarely warranted in normal situations and, for manual calculations, an approximation may be employed which eliminates the lengthy relaxation process which results from the closed loops of the frame. This is Bolton's substitute frame technique which is described in the following section.

4.5 Elastic Critical Loads of No-Sway Frameworks: Bolton's Substitute Frame Technique

The basis of this approach consists of testing a substitute frame which, as far as the tested joint is concerned, behaves in a similar manner to the real frame as it is argued that the stability of a joint is not significantly affected by members more than two joints removed. As with Winter's method it is necessary to first calculate the relative axial loads, Euler loads and relative stiffnesses of all members and then to identify the 'weakest' joint. Now, however, to obviate the iterative moment distribution part of the calculation a substitute frame is constructed as follows. Consider again the frame of Fig. 4.8. Suppose that joint

D is identified as the 'weakest' joint. Draw in all members meeting at the joint to be tested (D). At the ends of each of these members remote from D draw in all members meeting at the joint, but make the far ends of these members encastré. This means that several members will be drawn in twice as indicated in Fig. 4.13(a) but note that there are now no closed loops in the frame. Had joint E been selected for test the resulting substitute frame would be as indicated in Fig. 4.13(b).

As the tested joint, D, is rotated through unit angle a moment T_D (equal to Σks) appears at D. Of this moment, $k_{iD}s_{iD}$ will arise on member iD where i is any joint adjacent to D and a moment of $k_{iD}s_{iD}c_{iD}$ will be carried over to i. When joint i is balanced the magnitude of the balancing moment on iD will be $k_{iD}^2 s_{iD}^2 c_{iD}/T_i$, where $T_i = \Sigma_i ks$ for the summation being taken over all members meeting at i; therefore in turn a moment $-(k_{id}s_{iD}c_{iD})^2/T_i$ will be returned to D. Obviously due to the construction of the substitute frame all other carryover terms emanating from joint i will be absorbed into the fixed ends of all other members framing into i, and will no longer affect subsequent calculations. Thus one cycle of operations is sufficient to balance all joints other than the one under test and the moment appearing here will be

$$M_D = T_D - \Sigma_{iD} \frac{(k_{iD}s_{iD}c_{iD})^2}{S_i} \qquad (4.6)$$

where the summation Σ_{iD} is taken over all members meeting at D. This moment is a measure of the stiffness of the substitute frame. If M_D is positive then the frame is stable and if it is negative then the frame is unstable. To find the elastic critical load it is necessary to evaluate the moment for several trial values of the axial load parameter in order to determine when the moment changes sign, i.e. when the stiffness vanishes.

The following example illustrates the method and enables comparisons of time and accuracy to be made with Winter's method.

Example 4.IV

Consider again the rigidly jointed framework of Fig. 4.8 with the properties listed in Example 4.III.

Steps 1 to 4   These are carried out and tabulated as in Fig. 4.9 for the previous example.
Step 5   Again a trial value of the loading must be adopted and a value of ρ in DE of $2 \cdot 00$ is selected. Values of s and ks may be determined for all members

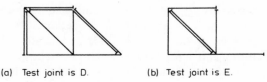

(a) Test joint is D.   (b) Test joint is E.

Fig. 4.13

Member	Test ρ	s	ks	$(sc)^2$	$(ksc)^2$	$\dfrac{(ksc)^2}{T_F}$
AB	$-1 \cdot 00$	$5 \cdot 18$	$5 \cdot 18$			
AD	$-2 \cdot 83$	$6 \cdot 85$	$4 \cdot 85$	$2 \cdot 29$	$1 \cdot 15$	$0 \cdot 09$
AE	$1 \cdot 00$	$2 \cdot 47$	$2 \cdot 47$			
BC	$-2 \cdot 83$	$6 \cdot 85$	$4 \cdot 85$			
BD	$1 \cdot 00$	$2 \cdot 47$	$2 \cdot 47$	$6 \cdot 09$	$6 \cdot 09$	$0 \cdot 49$
CD	$1 \cdot 00$	$2 \cdot 47$	$2 \cdot 47$	$6 \cdot 09$	$6 \cdot 09$	$0 \cdot 83$
DE	$2 \cdot 00$	$0 \cdot 143$	$0 \cdot 143$	$12 \cdot 42$	$12 \cdot 42$	$4 \cdot 76$
						$\Sigma\ 6 \cdot 17$

$T_A = 12 \cdot 49$ $T_B = 12 \cdot 49$ $T_C = 7 \cdot 31$ $T_D = 9 \cdot 99$ $T_E = 2 \cdot 61$

$\Sigma(ksc)^2/T_F < T_D$ $\therefore$ Stable $\Sigma/T_D = 0 \cdot 618$

Fig. 4.14

and $T = \Sigma ks$ may be evaluated for all joints in the substitute frame which can rotate. Values of $(sc)^2$ are tabulated and hence $(ksc)^2$ can be found. These values are required only for members which meet at the tested joint. Dividing by the stiffness of the joint at the 'far end' of the member (T_F) gives the entries in the final column of Fig. 4.14 and the summation of these quantities is a measure of the negative moment carried back to the tested joint after the single cycle of balancing operations which occurs with this method. In this instance $\Sigma(ksc)^2/T_F$ is $6 \cdot 17$ units which is less than T_D which is $9 \cdot 99$ units. Therefore the frame is stable at this value of load. The ratio of $\Sigma(ksc)^2/T_F$ to T_D is evaluated as this is a convenient measure of the stiffness of the frame — the frame has zero stiffness when this quantity is unity.

Step 6   Increase (or decrease) the value of the load parameter and repeat Step 5 as necessary. For example, if values of ρ of $2 \cdot 20$ and $2 \cdot 12$ in member DE are tested, the values of Σ/T which result are $1 \cdot 117$ and $0 \cdot 857$. These may be plotted as in Fig. 4.15 from which it can be seen that the critical value of ρ in

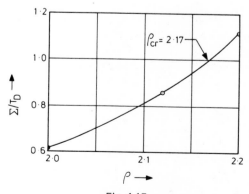

Fig. 4.15

member DE is about 2·17 compared with 2·16 found using Winter's relaxation approach. The evaluation of W_{cr} is carried out as at the end of the previous example.

4.6. Elastic Critical Loads of No-Sway Frameworks: Braced Portals

The foregoing methods can be adopted for any rigidly jointed no-sway frame provided reasonable estimates for the axial loads in the members can be made. For example the elastic critical load of the simple symmetrical portal frame of Fig. 4.16 subject to symmetrical loading may be readily obtained. Although the light bracing shown prevents sway in the frame, its contribution to the rotational stiffness of the joints will be considered insignificant and is therefore neglected. Intuitively the critical mode of this frame can be seen to be one in which B and C rotate through equal angles but in opposite senses as shown in Fig. 4.17. The EI/L values for the column and the beam are denoted by k_C and k_B respectively and the moments arising in the frame due to clockwise and anticlockwise rotations of θ at B and C are shown on Fig. 4.17.

The elastic critical load is found by equating the total moment at joint B or joint C to zero.

i.e.

$$sk_C + 2k_B = 0$$

or

$$s = -2k_B/k_C$$

for $k_B = k_C$ the stability criterion becomes

$$s = -2·00$$

and the corresponding value of ρ is given by

$$\rho_{cr} = 2·55$$

Thus

$$W_{cr} = 2·55 \, \pi^2 \, (EI/L^2)_{AB}$$

For

$$k_B = 2 \, k_C \text{ the instability criterion is}$$

$$s = -4·00$$

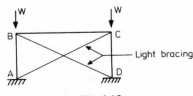

Fig. 4.16

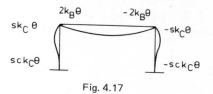

Fig. 4.17

giving

$$\rho_{cr} = 2 \cdot 88$$

and

$$W_{cr} = 2 \cdot 88\ \pi^2\ (EI/L^2)_{AB}$$

If the feet of the stanchions are considered to be pinned against in-plane rotations as shown in Fig. 4.18 the instability criterion would become

$$s''k_C + 2k_B = 0$$

For $k_B = k_C$ this reduces to

$$s'' = -2 \cdot 00$$

giving

$$\rho_{cr} = 1 \cdot 31$$

and

$$W_{cr} = 1 \cdot 31\ \pi^2\ (EI/L^2)_{AB}$$

For $k_B = 2k_C$ the instability criterion is

$$s'' = -4 \cdot 00$$

giving

$$\rho_{cr} = 1 \cdot 49$$

and

$$W_{cr} = 1 \cdot 49\ \pi^2\ (EI/L^2)_{AB}$$

Comparison with similar unbraced frames is made in Fig. 4.32.

The critical load charts of the Joint Committee's Second Report on fully-rigid multistorey welded steel frames produced by the Institution of Structural Engineers and the Welding Institute may readily be derived using this type of approach as developed by Wood. Considering the design of one column it is

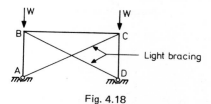

Fig. 4.18

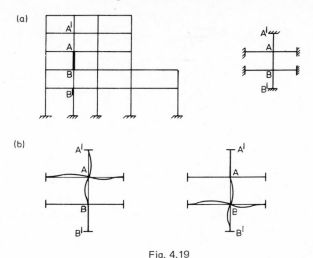

Fig. 4.19

assumed that the remote ends of all members attached to the ends of the column under consideration are fixed as in Bolton's substitute frame. It is further assumed that any columns above and below the column considered have identical I/L and ρ values to the column.

Consider an internal column AB of a rigidly jointed multistorey frame which is braced against sway (see Fig. 4.19(a)). The column is isolated together with the other members which frame into A and B. The columns AA^1 and BB^1 are assumed to be identical to AB and all joints other than A and B are assumed to be encastré. As this is a no-sway frame only two components of distortion are possible – a rotation of A and a rotation of B as shown in Fig. 4.19(b). The total moments which arise at A and B due to these distortions are

$$M_A = (2sk_C + 4k_{BT})\theta_A \quad M_B = sck_C\theta_A$$
$$M_A = sck_C\theta_B \quad\quad M_B = (2sk_C + 4k_{BB})\theta_B$$

where k_C is EI/L for the columns

$\quad$ k_{BT} is the sum of the EI/L values of the beams framing into the top joint A

and

$\quad$ k_{BB} is the sum of the EI/L values of the beams framing into the bottom joint B.

Eliminating the moment at A (or the moment at B) and putting the stiffness of the other joint to zero, the instability criterion is readily obtained as

$$(2sk_C + 4k_{BT})(2sk_C + 4k_{BB}) - (sc)^2 k_C^2 = 0$$

Dividing by k_C^2 gives

$$\left\{2s + \frac{4k_{BT}}{k_C}\right\}\left\{2s + \frac{4k_{BB}}{k_C}\right\} - (sc)^2 = 0$$

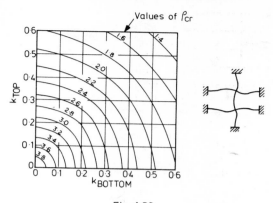

Fig. 4.20

Now it is convenient to use values of the normal Hardy Cross distribution coefficients k_{Top} and k_{Bottom} as basic variables as these will normally have been evaluated during the calculation of bending moments in the frame.

k_{Top} is defined as $k_C/\Sigma_A k = k_C/(k_{BT} + 2k_C)$

and

k_{Bottom} is defined as $k_C/\Sigma_B k = k_C/(k_{BB} + 2k_C)$

Equation 4.6 can be modified as follows:

$$\left\{2s + \frac{4k_{BT} + 8k_C}{k_C} - 8\right\} \left\{2s + \frac{4k_{BB} + 8k_C}{k_C} - 8\right\} - (sc)^2 = 0$$

or

$$\left\{2s + \frac{4}{k_{Top}} - 8\right\} \left\{2s + \frac{4}{k_{Bottom}} - 8\right\} - (sc)^2 = 0 \qquad (4.7)$$

This equation can be solved by a trial and error process as in the example below and the contour plot of Fig. 4.20 produced.

Example 4.V

For $k_{Top} = k_{Bottom} = 0.2$. Equation 4.7 becomes

$$(2s + 12)^2 - (sc)^2 = 0$$

or

$$2s + 12 \pm sc = 0$$

This equation is solved in the table of Fig. 4.21 from which it can be seen that $\rho = 2.76$ is the lowest value of axial load parameter for which Eq. 4.7 is satisfied and this corresponds to the elastic critical load of the substitute frame.

It should here be noted that the assumption regarding full fixity at the ends of the beams might appear unduly conservative as the beams will normally

	Trial 1	Trial 2	Trial 3	Trial 4	Trial 5	Trial 6	Trial 7
ρ	0	1·0	2·0	3·0	2·8	2·7	2·76
s	4	2·47	0·14	−5·03	−3·44	−2·09	−3·18
(2s + 12)	20	16·9	12·3	1·94	5·11	7·83	5·64
sc	2	2·5	3·5	7·12	5·88	5·42	5·69
(2s + 12 ± sc)	18	14·4	8·8	−5·18	−0·77	2·41	−0·05

Fig. 4.21

undergo single curvature. This is considered to be admissible due to composite beam/slab interaction which substantially increases the operative beam stiffness. If doubts exist about this assumption for a particular situation it can be removed by assuming the remote ends of the beam to be pinned and replacing the beam stiffness by an effective stiffness equal to three-quarters of the actual value.

Although this approach contains approximation it does represent a significant improvement on the crude stability analysis inherent in the estimation of effective lengths. If the initial assumption that all three columns in the substitute frame are very similar is not justifiable in a particular situation then the basic analysis may be carried out in order to obtain a close assessment of ρ_{cr}, or reference may be made to Wood's stiffness distribution method discussed in the following section.

4.7 Wood's 'Stiffness Distribution' Approach

A development of this work is Wood's concept of stiffness distribution. Consider a column AB, pinned at its upper end and attached to beams of stiffness k_{BB} at its lower end in a sway-braced frame as indicated in Fig. 4.22. If end A is rotated through θ then a moment of $sk\theta$ will arise at A and a moment of $sck\theta$ will be carried over to B. If a distribution coefficient DC_{BA} is defined as the moment arising on member AB at B when unit-balancing moment is applied at B in the presence of axial loads, then

$$DC_{BA} = s_{AB}k_{AB}/(s_{AB}k_{AB} + 4k_{BB}) \qquad (4.8)$$

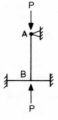

Fig. 4.22

Thus in the balancing operation at B the value of the balancing moment arising on AB at B, M_{balBA}, is given by

$$M_{balBA} = - sck\theta \times DC_{BA}$$

and the moment carried back to A, M_{balAB} is

$$M_{balAB} = - sc^2 k\theta \times DC_{BA}$$

Thus the total moment at A is

$$M_A = sk\theta - sc^2 k\theta \ DC_{BA}$$

$$= sk(1 - c^2 \times DC_{BA})$$

Or the modified stiffness, denoted by K'', is given by

$$K'' = \left\{ \begin{matrix} \text{stiffness of the member} \\ \text{assuming the remote} \\ \text{end fixed but including} \\ \text{the effect of axial load} \end{matrix} \right\} \left\{ 1 - \left[\begin{matrix} \text{carry} \\ \text{over} \\ \text{factor} \end{matrix} \right]^2 \left[\begin{matrix} \text{distribution} \\ \text{coefficient} \\ \text{of remote end} \end{matrix} \right] \right\} \quad (4.9)$$

This result which has a marked similarity with ordinary moment distribution (both linear and non-linear) gives an effective stiffness. It can only be evaluated numerically at specific values of the axial load parameter but these values may then be employed to determine whether or not a system has a positive (i.e. stable) or negative (i.e. unstable) stiffness. For example if end A were connected to beams with a total stiffness of k_{BT} then the net stiffness of joint A would be

$$K'' + 4k_{BT}$$

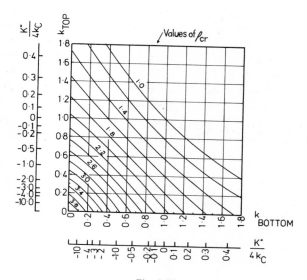

Fig. 4.23

Expanding for K'' and putting the resulting expression equal to zero for instability,

$$sk_C \left\{ 1 - c^2 \left[\frac{sk_C}{sk_C + 4k_{BB}} \right] \right\} + 4k_{BT} = 0 \qquad (4.10)$$

This expression will be found to be similar in form to Eq. 4.7 of Section 4.6 which treated a similar column/beam assemblage but which differed from the frame currently under consideration in that in the earlier example the column was assumed to be extended above and below the bottom joints. This feature accounts for the terms in 2s in Eq. 4.6 which correspond to the terms in s in Eq. 4.10.

The general solution of Eq. 4.10 can be presented in chart form as shown in Fig. 4.23 which is similar in appearance to Fig. 4.20 but variations in values occur as a result of the differences in the respective frames under consideration. The latter version is more general as the next example illustrates.

Example 4.VI

Determine the elastic critical load of the limited frame of Example 4.V using stiffness distribution.

Now

$$k_{TOP} = 0.2 = k_{BOTTOM}$$

But

$$k_{TOP} = k_C / (k_{BT} + 2k_C)$$

or

$$k_{BT} = 3k_C$$

Similarly

$$k_{BB} = 3k_C$$

These values of k_{BB} and k_{BT} would normally be input parameters from which k_{TOP} and k_{BOTTOM} would be evaluated. Guess a trial value of $\rho = 2.50$ then $s = -1.75$ and $c = -2.67$. Consider first the bottom joint; referring to the definition of K'' the effective stiffness, K'', of the column AB at A is

$$-1.75 \times k_C \{ 1 - [(2.67)^2 (-1.75k_C)/(4 \times 3k_C - 2 \times 1.75_C)] \}$$
$$= -4.33k_C$$

The total stiffness of joint is therefore

$$4 \times 3k_C - 1.75k_C - 4.33k_C = +5.92k_C$$

This is positive and the system is stable at $\rho = 2.50$.
Try $\rho = 2.80$ giving $s = -3.44$ and $c = -1.71$.

The effective stiffness of AB at A is

$$-3.44k_C \{1 - [(1.71)^2(-3.44k_C)/(4 \times 3k_C - 2 \times 3.44k_C)]\} = 10.22k_C.$$

The total stiffness of joint A is therefore

$$4 \times 3k_C - 3.44k_C - 10.22k_C = -1.66k_C.$$

This is less than zero implying that the frame has a negative stiffness and the system is unstable at $\rho = 2.80$. Repeat calculations at $\rho = 2.70$ and $\rho = 2.76$ give total stiffnesses of joint A of $1.79k_C$ and $-0.10k_C$ respectively giving $\rho_{cr} = 2.76$ which is in accordance with the solution developed in Example 4.V.

The chart of Fig. 4.23 may be employed in a different manner in order to evaluate the effective modified stiffness of a column with any remote end constraint using a successive stiffness distribution technique. This approach requires that additional axes be appended to the plot. In developing Eq. 4.10 use was made of the fact that the total stiffness of the joint under consideration was $K'' + 4k_{BT}$ and for instability this was zero. The plot of Fig. 4.23 was derived using this condition. The ratio of the modified column stiffness K'', to the full column stiffness in the absence of axial loads is then $K''/4k_C$ which is, at instability, given by

$$\frac{K''}{4k_C} = -\frac{k_{BT}}{k_C}$$

But

$$k_{TOP} = k_C/(k_C + k_{BT})$$

which gives

$$-\frac{k_{BT}}{k_C} = +\left\{1 - \frac{1}{k_{TOP}}\right\} = \frac{K''}{4k_C} \tag{4.11}$$

This implies that a new scale may be determined for Fig. 4.23 by simply substituting values for k_{TOP} into Eq. 4.11, e.g. when $k_{TOP} = 0.2$; $K''/4k_C = 1 - (1/0.2) = -4$. A similar substitution may be made for the other axis.

The plot can now be used to determine the effective column stiffness for a specific trial value of axial load parameter if the Hardy Cross distribution coefficient at one end of the column is known. For example if $k_{BOTTOM} = 0.6$ and $\rho_{cr} = 2.0$ then $K''/4k_C = -1.1$. Successive usage of this approach means that effective lengths of columns in extensive frames may be found. This may be warranted in a frame where it appears that several column lifts may become unstable at about the same load and each may significantly interact with restraining beams and more stable column lengths. The following example illustrates the use of the method which may be readily extended to deal with any number of storeys.

Example 4.VII

Determine the elastic critical load of the limited frame of Example 4.V using the chart of Fig. 4.23.

For this frame:

$$k_{BT} = k_{BB} = 3k_C$$

$$k_{upper} = k_{lower} = k_C$$

The actual computation is best carried out in tabular form as in Fig. 4.24. Select first a trial value of ρ of $2 \cdot 50$. (In this example ρ has the same value throughout all three storeys.) This does not imply a limitation on this stiffness distribution method although it was a necessary limitation in the derivation of Fig. 4.20 which was used in Example 4.V.) Starting at the top of the frame, the upper end of the upper lift is fixed giving $k_{TOP} = 0$. Entering Fig. 4.23 with $k_{TOP} = 0$ and $\rho = 2 \cdot 50$ gives

$$K''/4k = 0 \cdot 45.$$

Thus for the central lift the effective restraint at the top of the column is $K'' + 4k_B$ or

$$4 \times 3k_C - 4 \times 0 \cdot 45k_C = 4 \times 2 \cdot 55k_C.$$

The nominal distribution coefficient may now be evaluated as

$$4k_C/(4k_C + K'') = 4k_C/(4k_C + 4 \times 2 \cdot 55k_C) = 0 \cdot 282.$$

With this figure and the value of ρ in the central column AB, Fig. 4.23 may be reused to evaluate the effective stiffness of the column AB at B (including the effects of the restraint at A) which is found to be $-1 \cdot 10$.

Test ρ		2·50	2·60	2·70	2·80
upper column	k_t	0	0	0	0
	ρ	2·50	2·60	2·70	2·80
	$K''/4k_C$	−0·45	−0·51	−0·70	−0·86
k_{BT}		3·00	3·00	3·00	3·00
effective beam stiffness		$4 \times 3 - 4 \times 0 \cdot 45$ $= 4 \times 2 \cdot 55$	4 × 2·49	4 × 2·30	4 × 2·14
central column	k_t	0·282	0·286	0·303	0·318
	ρ	2·50	2·60	2·70	2·80
	$K''/4k_C$	−1·10	−1·45	−1·80	−2·60
k_{BB}		3·00	3·00	3·00	3·00
lower column	$K''/4k_C$	−0·45	−0·51	−0·70	−0·86
	ρ	2·50	2·60	2·70	2·80
	k_b	0	0	0	0
Net stiffness at B		4 × 1·45	4 × 1·04	4 × 0·50	4 × −0·46

Fig. 4.24

Working from the bottom end of the structure, the lower end of the column is fixed giving $k_{BOTTOM} = 0$ and as $\rho = 2 \cdot 50$ in this column also, the effective stiffness of this column at B may be found to be $-0 \cdot 45$ from Fig. 4.23. At B the net stiffness can now be determined as

$$4 \times k_{BB} + K''_{AB} + K''_{lower}$$

i.e.

$$4 \times 3k_C - 4 \times 1 \cdot 1 k_C - 4 \times 0 \cdot 45 k_C = 4 \times 1 \cdot 45 k_C$$

Clearly the factors of 4 may be omitted from the calculations (as happens in Wood's paper where a new basic stiffness of $S/4$ is defined) but it is retained here for consistency with earlier work. Similar calculations at $\rho = 2 \cdot 60$, $2 \cdot 70$ and $2 \cdot 80$ show a falling stiffness of joint B which becomes zero between $\rho = 2 \cdot 70$ and $2 \cdot 80$ which gives the critical value of P/P_E. Interpolation of the stiffnesses shows this value to be $2 \cdot 76$ as obtained in Examples 4.V and 4.VI.

All of the methods discussed so far in this chapter are applicable only to frames where sway is prevented, either by the triangulated nature of the frame or by bracing or shear walls if the frame is of portal frame construction. If sway is not prevented then additional modes of distortion must be considered.

4.8 Elastic Critical Loads of Sway Frameworks: No-Shear Functions n and o

The patterns of moments technique, Winter's method and Bolton's substitute frame approach may all be suitably modified to deal with frames in this category. The simplest form of portal frame which can be considered is a single-storey symmetrical frame subject to symmetrical loading such as that shown in Fig. 4.25. Despite the symmetry of both the frame and the loading, the antisymmetrical sway mode of distortion is possible and will be seen to occur at a load very much less than that associated with the symmetrical mode (see the braced portal of Section 4.6). This sway mode is made up of two basic components of distortion, the first equal and like rotations of B and C, and the second a sway. The patterns of moments corresponding to these two distortions are shown in Fig. 4.26, together with the additional constraint which must be considered, namely the propping force at beam level. For the purely rotational pattern (where the sway of the columns is zero) this is the sum of the column end moments divided by the storey height, whilst for the pure sway mode the same expression must be modified by the sway factor m as derived in Chapter 2. Thus the problem contains two disturbances (the rotations and the sway) and

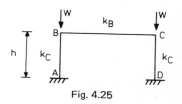

Fig. 4.25

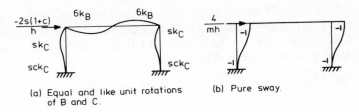

(a) Equal and like unit rotations of B and C.　　　(b) Pure sway.

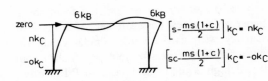

(c) Unit "no-shear" rotation of B and C.

Fig. 4.26

two constraints (the moments and the propping force). It is convenient to eliminate the propping force by combining the patterns as that of Fig. 4.26(a) plus $ms(1 + c)k_C/2$ parts of the pattern of Fig. 4.26(b). The resulting pattern is shown in Fig. 4.26(c) from which it can be seen that the terms in the brackets are dependent solely upon the axial load parameter ρ. The moment in the column at beam level is put equal to $n\,k_C$ and that at the base of the column is

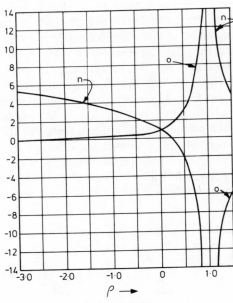

Fig. 4.27

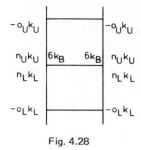

Fig. 4.28

put equal to $-o\,k_C$: i.e.

$$n = [s - ms(1 + c)/2] \tag{4.12}$$

and

$$o = -[sc - ms(1 + c)/2] \tag{4.13}$$

These functions were first defined by Merchant and are known as 'no-shear' functions. They are listed in the tables of stability functions and the way in which they vary can be seen from Fig. 4.27. (It will be noticed that when ρ is zero both n and o are equal to unity.) The distortion corresponding to the situation of Fig. 4.26(c) is one in which the joints at the beam level have equal and like unit rotations and there is sufficient sway of the storey so that no external propping force is required for equilibrium. The pattern of moments for this distortion may be written down as $6k_B$ on the beams together with nk_C and $-ok_C$ at the 'near' and 'other' ends of the columns. The correspondence to Naylor's no-shear distribution factors is apparent if it is remembered that in the absence of axial loads n and o both have the value of unity.

The no-shear pattern of moments for an intermediate floor in a single-bay multistorey frame is indicated in Fig. 4.28 where the suffixes 'L' and 'U' relate to the lower and upper columns respectively.

4.9 Elastic Critical Loads of Sway Frameworks: Patterns of Moments Technique

Considering the frame of Fig. 4.25, the elastic critical load may be found by simple consideration of the no-shear distortion pattern of Fig. 4.26(c). The total moment appearing at each beam/column joint is given by

$$M = 6k_B + nk_C$$

and the condition for instability is that this quantity is zero:

$$M = 6k_B + nk_C = 0 \tag{4.14}$$

For the special case of $k_B = k_C$ Eq. 4.14 reduces to

$$n = -6$$

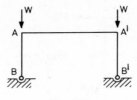

Fig. 4.29

From Fig. 4.27 it can be seen that this first occurs when

$$\rho = 0{\cdot}748$$

Thus

$$W_{cr} = 0{\cdot}748 \; \pi^2 \; (EI/L^2)_{AB}$$

Note the drop in critical load from $2{\cdot}55 \; P_{EAB}$ for the same frame with sway bracing considered in Section 4.6. For $k_B = 2k_C$ the condition of instability is

$$n = -12$$

giving

$$\rho = 0{\cdot}854$$

and

$$W_{cr} = 0{\cdot}854 \; \pi^2 \; (EI/L^2)_{AB}$$

Again comparison with the corresponding sway-braced frame of Section 4.6 reveals a fall in elastic critical load from $2{\cdot}88 \; P_{EAB}$ to $0{\cdot}854 \; P_{EAB}$.

If the feet of the columns are considered to be pinned against in-plane rotations as shown in Fig. 4.29 then it would be necessary to permit A and B to rotate until the out-of-balance moments appearing at these joints disappeared. If a normal rotation pattern were employed to do this a propping force would become necessary once more. Therefore these moments must be balanced by no-shear rotations of B and B$'$ as in Fig. 4.30(b). Combining these two patterns in the correct proportions (by taking o/n parts of (b) together with (a)) gives the pattern of moments shown in Fig. 4.31 which satisfies all of the necessary conditions.

The total moment at A or A$'$, which is a measure of the stiffness of the frame when subjected to equal rotational disturbances without any lateral restraint, is given by

$$M_B = M_C = 6k_B + (n - o^2/n)k_C$$

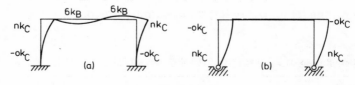

Fig. 4.30

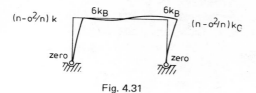

Fig. 4.31

For elastic instability this stiffness vanishes and the resulting equation (4.15) must be solved by trial and error.

$$6k_B + (n - o^2/n)k_C = 0 \qquad\qquad (4.15)$$

If $k_B = k_C$ this equation reduces to

$$o^2/n = 6 + n$$

giving

$$\rho_{cr} = 0 \cdot 184$$

and

$$W_{cr} = 0 \cdot 184 \, \pi^2 (EI/L^2)_{AB}$$

If $k_B = 2k_C$ Eq. 4.15 reduces to

$$o^2/n = 12 + n$$

giving

$$\rho = 0 \cdot 213$$

and

$$W_{cr} = 0 \cdot 213 \pi^2 (EI/L^2)_{AB}$$

Comparison of all results for sway-braced and unbraced sway frames with fixed and pinned feet is made in Fig. 4.32.

This approach, namely the combination of basic patterns of distortion, can be employed on frames up to about three storeys in height. Comparison with the patterns of moments technique for triangulated frames should be made.

	Values of ρ_{cr}			
	No-Sway Frames		Sway Frames	
	$k_B = k_C$	$k_B = 2k_C$	$k_B = k_C$	$k_B = 2k_C$
Feet Fixed	2·55	2·88	0·748	0·854
Feet Pinned	1·31	1·49	0·184	0·213

Fig. 4.32

For taller structures however this approach becomes too unwieldy and an alternative is preferable. Three such approaches are described in subsequent sections of this chapter; each of these produces results of varying degrees of precision and requires greatly varying time and effort.

4.10 Elastic Critical Loads of Sway Frameworks: Winter's Relaxation Technique

This method is basically similar to that developed by Winter for triangulated frameworks. Its application is limited essentially to symmetrical single bay frames but if a close approximation to the elastic critical load will suffice then a multibay frame can be transformed into an equivalent single-bay framework as is described in standard texts on moment distribution. The total loading at instability of the equivalent frame may then be found from which the value of λ_{cr} for the real frame may be determined.

In the form in which the method is employed for triangulated frameworks, distribution coefficients (e.g. $DC_{i \to j}$) are calculated. These relate the moment carried over to an adjacent joint, j, when unit moment is balanced by pure rotation at i. For sway frames a modification is required so that the shear equilibrium is preserved, i.e. so that no external propping forces are required at floor levels. To accomplish this the pure rotational balancing operations are replaced by 'no-shear' balancing operations, thus ensuring that shear equilibrium is automatically maintained.

Consider the balancing of any pair of joints I, I' in a single bay multistorey sway frame. The pattern of moments corresponding to a no-shear rotation of θ at I and I' is as shown in Fig. 4.33. If this pattern of moments is being employed to apply unit balancing moment at I and I' then the moment carried over to the adjacent joints J and J' (known as the distribution coefficient $DC_{I \to J}$) is given by

$$DC_{I \to J} = \frac{-k_{IJ}o_{IJ}}{6k_{II'} + k_{HI}n_{HI} + k_{IJ}n_{IJ}} \tag{4.16}$$

This is identical to $DC_{I' \to J'}$ and clearly only one column in the frame need be considered in the computation.

Unit rotation could now be applied to both columns at any beam level and no-shear moment distribution carried out until the residuals are sufficiently

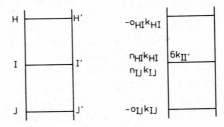

Fig. 4.33

small enabling a numerical measure of the stiffness of the frame to be determined. Because of the nature of the frame and the magnitudes of the distribution coefficients, convergence is likely to be very slow and many cycles of balancing are required before convergence is obtained and a more rapid test is therefore desirable. The process adopted is to determine whether or not convergence will occur, i.e. to establish whether or not a set of moments exists which is in equilibrium with the distortion applied. This is accomplished by examining the balancing moments after each cycle of carryover operations. If the balancing moments at every joint on the column are all less than any previous set then it is concluded that eventually the moments will converge to a finite distribution and the frame is stable. If however these balancing moments are all greater than any previous set, then it is concluded that the residuals are growing and the moments will not converge to a finite distribution and the frame is unstable. The process is repeated for several trial values of axial load parameter until sufficiently close upper and lower bounds to the critical load are established. (At the critical load the moments neither converge nor diverge.) Interpolation to determine a closer estimate of ρ_{cr} is not possible as a numerical measure of the frame stiffness has not been found.

An improvement which leads to a further speeding-up of the convergence is to apply a test distortion which involves no-shear rotations of all joints. The closer this initial pattern of test distortions is to the critical mode, the more rapid will be the convergence. In the simple 'artificial' example which follows it will be noticed that equal moments are applied to all joints except the top one where only one-half of the value is used. This has been done very simply by applying equal moments to the ends of every column. The distribution of the initial test moments does not affect the value of the elastic critical load which is computed; it does however influence the convergence process and therefore the quantity of numerical effort necessary to establish convergence or divergence.

If the distortion pattern corresponding to the applied test differs significantly from the elastic critical mode then it may seem from the first few cycles of balancing operations that convergence will occur, i.e. the majority of the joints are associated with falling residuals. It is incorrect however to assume that the frame is stable at the test load, unless the moments at *all* joints are less than those in a previous line. This is because, even if the test load is greater than the elastic critical load, all of the test distortion which is not in the critical mode is convergent and will, after successive cycles of balancing, disappear. That part of the test distortion which is in the critical mode, however, will be magnified and ultimately divergence will be established. Thus the first few cycles of operations may be inconclusive. If it were possible to write the test moments down in the ratios corresponding to the critical mode, only one cycle of balancing would be necessary to determine convergence or divergence.

It is important to note that, when the distribution coefficients $DC_{I \to J}$ are determined from Eq. 4.16, the denominator must be positive. Clearly when no axial loads are present the denominator is positive indicating a finite stiffness for joints I and I' assuming H, H', J and J' are rigidly constrained against rotation. If a trial value of axial load is selected which results in the denominator being zero

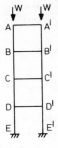

Fig. 4.34

(or even negative) then this implies that II' would be unstable even if those parts of the frame restraining H, H' and J and J' were infinitely stiff. This trial value of axial load parameter must be reduced. Note that when the denominator is zero the distribution coefficient will be infinity with divergence resulting in the subsequent analysis.

The use of the method is demonstrated by a very simple example. This simplicity is deliberate as the example is designed so that the basis of the method may be more readily appreciated. This does not however imply a limitation to the method.

Example 4. VIII

Consider the simple frame of Fig. 4.34 in which all the members are identical and equal loads are applied at the top of each column.

Step 1   Determine relative values of k for all members and relative values of ρ for each stanchion. (Hence all k values are the same and all ρ values are identical.)

Step 2   Select some trial value of ρ. A value of $0 \cdot 42$ in the columns is chosen.

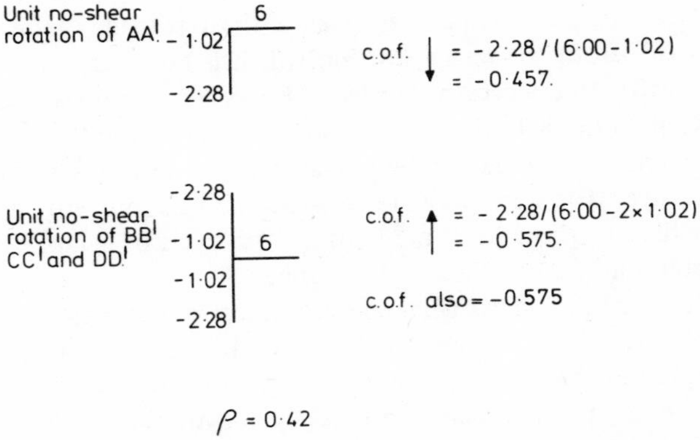

Unit no-shear
rotation of AA'

$-1 \cdot 02$ ⌐———6

$-2 \cdot 28$

c.o.f. ↓ = $-2 \cdot 28 / (6 \cdot 00 - 1 \cdot 02)$
= $-0 \cdot 457$.

$-2 \cdot 28$

Unit no-shear
rotation of BB'
CC' and DD' $-1 \cdot 02$ | 6

$-1 \cdot 02$

$-2 \cdot 28$

c.o.f. ↑ = $-2 \cdot 28 / (6 \cdot 00 - 2 \times 1 \cdot 02)$
= $-0 \cdot 575$.

c.o.f. also = $-0 \cdot 575$

$\rho = 0 \cdot 42$

Fig. 4.35

	Bending moments at				
	D	C	B	A	Σ B.M.
Test moments	−20	−20		−20	−10	
Balance (1)	20	20		20	10	70
C.O.	−11·5	−11·5 −11·5	−11·5 −4·57	−11·50	
Balance (2)	11·5	23·0		16·07	11·50	62·1
C.O.	−13·22	−6·61 −9·24	−13·22 −5·26	−9·24	
Balance (3)	13·22	15·85		18·48	9·24	56·8

Fig. 4.36

Step 3   Consider no-shear rotations of the joints at each floor level (together with the appropriate sway of the storey immediately above and immediately below the level under consideration as shown in Fig. 4.33). At $\rho = 0·42$; $n = −1·02$ and $o = 2·28$. When joints A and A′ undergo unit no-shear distortion $(6 − 1·02) = 4·98$ units of moment arise at A and A′; moments of $−2·28$ units are caused at B and B′ and hence there is a carryover factor of $−2·28/4·98 = −0·457$ from A to B (and A′ to B′) as indicated in Fig. 4.35. Similarly applying no-shear distortions to any pair of joints BB′, CC′ or DD′ and their neighbouring stanchions, carryover factors equal to $−2·28/3·96 = −0·575$ are found.

A test disturbance which rotates one pair of joints in a no-shear manner could now be employed but a more rapid test is made by applying a pattern of moments throughout the frame as shown in the line labelled 'test moments' in Fig. 4.36. These moments must be balanced, followed by carryover operations to adjacent joints using the carryover factors calculated in Step 3. The total moment carried over to each joint must now be balanced followed by further carryover operations. The process is repeated until comparison of moments in balance (1) and balance (3) reveals that all of the moments in the latter are less than the corresponding entries in the former. It is therefore inferred that the moments are converging to finite values and thus the frame is stable at $\rho = 0·42$.

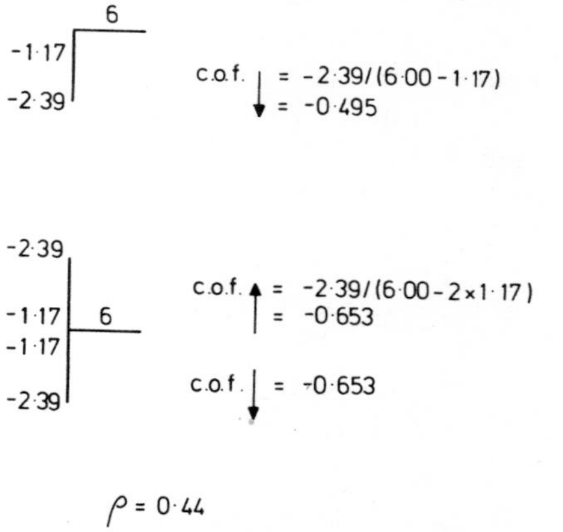

$\rho = 0·44$

Fig. 4.37

	Bending moments at				Σ B.M.
	D	C	B	A	
Test moments	−20	−20	−20	−10	
Balance (1)	20	20	20	10	70
C.O.		−13·06 −13·06 −13·06 −13·06 −4·95		−13·06	
Balance (2)	13·06	26·12	18·01	13·06	70·25
C.O.		−17·06 −8·53 −11·76 −17·06 −6·46		−11·76	
Balance (3)	17·06	20·29	23·52	11·76	72·63
C.O.		−13·25 −11·14 −15·36 −13·25 −5·82		−15·36	
Balance (4)	13·25	26·50	19·07	15·36	74·18

Fig. 4.38

Step 4   Increase the trial value of ρ and repeat Step 3. Test with $\rho = 0\cdot44$, i.e. $n = -1\cdot17$ and $o = 2\cdot39$. The no-shear patterns of moments are shown in Fig. 4.37 together with the calculation of the carryover factors. The Table of Fig. 4.38 shows that all moments are increasing and therefore it can be seen that the moments are diverging and the frame is unstable.

Step 5   Determine ρ_{cr}. As the frame is stable when $\rho = 0\cdot42$ and unstable when $\rho = 0\cdot44$ it may be concluded that $W_{cr} = 0\cdot43 \pm 0\cdot01$ times the Euler load of any column.

4.11 Elastic Critical Loads of Sway Frameworks: Wood's Method

The method adopted here is basically very similar to that developed by Wood for no-sway frames (see Section 4.6) in that the stability of the member under consideration is determined by making certain simplifications to the structure. Consider a single column rigidly attached to beams at either end in a frame where sway bracing is absent as indicated in Fig. 4.39. A convenient test for stability is a rotation of one end of the column, say the upper end. As sway bracing is absent the frame is unable to resist lateral forces and it is therefore appropriate to use no-shear distortions when considering the patterns of moments corresponding to the rotations of both ends of the column as indicated in Fig. 4.40. Combining these to eliminate net moment at the lower end leads to

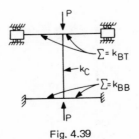

Fig. 4.39

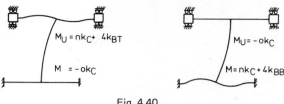

Fig. 4.40

the expression for M_U, which is the moment arising at the upper end of the column when unit rotation is applied there, the lower end being free to rotate and sway being allowed to occur until equilibrium is attained. M_U is a measure of the resistance of the frame which vanishes at instability.

$$M_U = 4k_{BT} + nk_C - \frac{o^2 k_C^2}{4k_{BB} + nk_C}$$

For instability

$$M_U = zero$$

or

$$(4k_{BT} + nk_C)(4k_{BB} + nk_C) - o^2 k_C^2 = 0 \qquad (4.17)$$

Once again it is convenient to use the normal Hardy-Cross distribution coefficients as input parameters. These are defined as:

$$k_{TOP} = k_C/(k_C + k_{BT})$$
$$k_{BOTTOM} = k_C/(k_C + k_{BB})$$

Equation 4.16 may be modified as follows:

$$\left\{ n + \frac{4k_{BT} + 4k_C}{k_C} - 4 \right\} \left\{ n + \frac{4k_{BB} + 4k_C}{k_C} - 4 \right\} - o^2 = 0$$

or

$$\left\{ n - 4 + \frac{4}{k_{TOP}} \right\} \left\{ n - 4 + \frac{4}{k_{BOTTOM}} \right\} - o^2 = 0 \qquad (4.18)$$

Equation 4.18 may be solved for combinations of k_{TOP} and k_{BOTTOM} to give the contour plot of Fig. 4.41 for the ratio P_{cr}/P_E. Remember that from Eq. 4.2 the effective length of the member may be determined

$$\ell = L/\sqrt{(P_{cr}/P_E)} \quad \text{or} \quad L/\sqrt{(\rho_{cr})}$$

Example 4.IX

Solve Eq. 4.18 for $k_{TOP} = k_{BOTTOM} = 0\cdot2$.

$$(n + 16)^2 - o^2 = 0$$

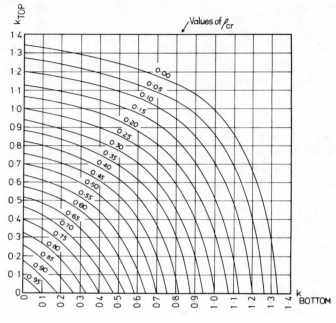

Fig. 4.41

or

$$(n + 16) \pm o = 0$$

This equation is solved in the Table of Fig. 4.42 from which it can be seen that ρ_{cr} is slightly above 0·79 with a corresponding effective column length of 1·12L.

Once the contour plot has been established values of ρ_{cr} may be read off directly without recourse to trial and error solutions and the usage of Fig. 4.41 is demonstrated in the next example.

This analysis assumes that the ends of the beam remote from the column do not rotate whereas when a sway frame becomes unstable the beams distort with

	Trial 1	Trial 2	Trial 3	Trial 4	Trial 5	Trial 6
ρ	0	0·4	0·6	0·8	0·7	0·79
n	1	−0·88	−2·84	−8·16	−4·66	−7·67
n + 16	17	15·12	13·16	7·84	11·34	8·33
o	1	2·17	3·74	8·63	5·35	8·16
n + 16 − o	16	12·95	9·42	−0·79	5·98	0·17

Fig. 4.42

double curvature thereby increasing the operative stiffness of the beams above that assumed. If equal rotations of the ends of the beams are assumed the operative beam stiffnesses are increased by 50% and this feature may be incorporated when evaluating the distribution coefficients. Clearly this cannot always represent a lower bound approach but composite beam/slab interaction will normally more than compensate for any non-conservatism.

Example 4.X

Consider the frame of Fig. 4.25 with $k_B = k_C$ which was found to have a value of W_{cr} of 0·748 times the Euler load of one column.

Due to symmetry both ends of the beam will rotate equally in the sway mode and therefore the operative stiffness of the beam will be exactly 50% greater than the actual value.

Let the column stiffness be k. Then the operative top beam stiffness is $1 \cdot 5k$. Bottom beam stiffness is infinite as the feet are assumed to be fully fixed.

$$k_{TOP} = k/(k + 1 \cdot 5k) = 0 \cdot 4$$

$$k_{BOTTOM} = k/(k + \infty) = 0$$

From Fig. 4.41 $P_{cr}/P_E = 0 \cdot 75$ (as before).

4.12 Elastic Critical Loads of Sway Frameworks: Horne's Method

In addition to the methods so far developed there is an alternative approach for multistorey multibay sway frames which should be very attractive to engineers because the only analysis which must be carried out is linear elastic. The method was developed using energy concepts employing the principle that, at the elastic critical load, a frame is in a state of neutral equilibrium and for small distortions there is no change in the total energy of the system. In this approach the loss of potential energy of the vertical loads is equated to the increase in the strain energy of the structure. The method is generally applicable but a major difficulty is that one is required to input the true critical mode shape in order to obtain the exact critical load. It can be shown that provided a mode shape, which is similar to the true form and which satisfies the boundary conditions, is used then good approximations to the critical load will be found, although these computed values will always exceed the true critical load, i.e. an unsafe solution will be found. The final expression derived using the approach briefly described above is exceedingly simple both in its form and in its usage.

It is however possible to obtain the result without the necessity of embarking on energy principles and, as this approach follows on from earlier sections and affords further insight into simple stability concepts, this process will be used here.

Consider the idealised structure shown in Fig. 4.43(a) which represents two adjacent floor levels in a multistorey building. The geometrical separation of the floors is provided by four rigid columns, freely pivoted at their ends. The spring

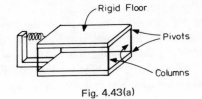

Fig. 4.43(a)

Fig. 4.43(b)

represents the stiffness of the columns to the sway distortion. The system may be further simplified into the arrangement of Fig. 4.43(b) in which the single rigid member AB is used to represent the columns. P is the total load carried by all columns at the floor level concerned. Under the influence of the vertical load P lateral motion of B will occur until equilibrium (if possible) is attained. It will be observed that two distinct actions are operating on the system namely:

(i) The vertical load P, acting at a sway eccentricity Δ, causing an overturning action about the base of the column. (This is frequently referred to as the P-Δ effect.)

(ii) The restoring force, F, due to the extension, Δ, of the spring which results in a restoring moment Fh about the base of the column.

When the load P is small the spring stiffness effect is larger than the P—Δ effect and equilibrium will be achieved, but if P is large then the P—Δ effect will dominate and the system will be unstable.

End A is pivoted and it is possible to take moments about A and observe that the net restoring moment M_R is given by

$$M_R = Fh - P\Delta \tag{4.19}$$

The spring force F may be written as SΔ where S is the stiffness of the spring and Eq. 4.19 then becomes

$$M_R = S\Delta h - P\Delta \tag{4.20}$$

If M_R is positive then the system will tend to revert to its initial undistorted position when slightly displaced but, when M_R is zero, the system will become unstable. Clearly this will occur when

$$P = Sh \tag{4.21}$$

This is more usually written as

$$\lambda_{cr}W = Sh \tag{4.22}$$

where W is the axial load in the columns under design conditions and λ_{cr} is the load factor at elastic instability.

Now S is the stiffness of the storey to a lateral disturbance and may be written as

$$S = H/u \tag{4.23}$$

where H is a horizontal disturbing force and u is the lateral deformation due to H.

If H is taken as W, substituting Eq. 4.23 into Eq. 4.22 gives

$$\lambda_{cr} = \frac{Sh}{W} = \frac{h}{u} = \frac{1}{\phi} \tag{4.24}$$

Thus the critical load factor is given by the reciprocal of the sway index ϕ.

For multistorey sway frames the lateral sway buckling load for any storey may be found as the reciprocal of the sway index for that storey under a total shear in the storey equal to the total axial load in the columns for that storey. This will give varying results throughout the height of the frame. However the overall in-plane stability of such frames can be found by monitoring the sway indices for all storeys and selecting the largest value to enable a close estimate for the elastic critical load to be found. The sway indices may be computed using normal linear elastic analyses and it is found most convenient to determine the sway indices simultaneously by applying to the frame a fictitious set of side loads at floor levels so that the shear in every storey is 1% of the total axial loads in the columns for that storey. Reducing H by a factor of one hundred results in deflections of reasonable magnitudes and also modifies Eq. 4.24 as indicated below:

$$\lambda_{cr} = \frac{Sh}{W} = \frac{h}{100u} = \frac{0 \cdot 01}{\phi} \tag{4.25}$$

The sway index ϕ should be calculated from an elastic analysis in which the rotational stiffnesses of the joints are reduced to allow for the presence of the axial load in the columns but this effect can normally be allowed for by reducing λ_{cr} by 10% giving

$$\lambda_{cr} = \frac{0 \cdot 009}{\phi} \tag{4.26}$$

The one exception to this is with single-storey frames where it is suggested that

$$\lambda_{cr} = \frac{0 \cdot 0083}{\phi} \tag{4.27}$$

be used.

These revisions ensure that the results obtained from Horne's approach are essentially conservative.

Example 4.XI

Consider the frame of Fig. 4.25 which was found to have a value of W_{cr} of 0·748 times the Euler load of one column when $k_B = k_C$.

A linear elastic analysis leads to an expression for the lateral deflection, δ, of the beam/column joint of

$$\delta = HL^3/16{\cdot}8EI$$

where H is the horizontal applied force at beam level.

Thus, if the design loading is W on the head of each column, putting

$$H = W/50 \text{ gives}$$

$$\delta = WL^3/840EI$$

Thus

$$\phi_{max} = WL^2/840EI$$

and

$$\lambda_{cr} = 0{\cdot}0083/\phi = 0{\cdot}0083 \times 840EI/WL^2$$

giving

$$\rho_{cr} = \lambda_{cr}W/P_E$$
$$= \frac{0{\cdot}0083 \times 840 \times EI \times W \times L^2}{WL^2 \times \pi^2 \times EI}$$
$$= 0{\cdot}706$$

compared with the exact solution of 0·748 determined earlier.

Example 4.XII

Consider again the frame of Fig. 4.34 with all members identical as in Example 4.VIII. Let L and EI denote the member lengths and flexural rigidities respectively. Using Horne's approach it is necessary to induce in each storey a horizontal shear equal to 1% of the total axial load in the columns in that storey. This is accomplished by applying a horizontal force of W/50 at roof level, i.e. at AA′.

A linear elastic analysis gives the deflections listed below from which sway indices for each storey may be readily computed as indicated.

Level	Deflection	Sway Index
AA′	$0{\cdot}00865WL^3/EI$	
		$\phi_{AB} = 0{\cdot}00216WL^2/EI$
BB′	$0{\cdot}00649WL^3/EI$	
		$\phi_{BC} = 0{\cdot}00249WL^2/EI$
CC′	$0{\cdot}00400WL^3/EI$	
		$\phi_{CD} = 0{\cdot}00242WL^2/EI$
DD′	$0{\cdot}00158WL^3/EI$	
		$\phi_{DE} = 0{\cdot}00158WL^2/EI$
EE′	$0{\cdot}00000$	

ϕ_{max} is identified as ϕ_{BC} which has a value of $0 \cdot 00249 WL^2 / EI$

$$\lambda_{cr} = 0 \cdot 009 / \phi_{max} = 0 \cdot 009 EI / 0 \cdot 00249 WL^2 = 3 \cdot 61 EI / WL^2 .$$

Now

$$\rho_{cr} = \lambda_{cr} W / P_E = \frac{3 \cdot 61 EI \; L^2}{L^2 \; \pi^2 EI} = 0 \cdot 366$$

This is a safe approximation to the value of $0 \cdot 43$ found using the more accurate method of Example 4.VIII.

4.13 Elastic Critical Loads and Frame Design

The new draft British Standard for Structural Steelwork differentiates between braced and sway frames in the treatment of overall frame instability. For braced frames it states that frame instability need not normally be considered. This implies that interaction is being neglected which represents a conservative approximation. The problem is then reduced essentially to one of column stability. In certain circumstances this leads to somewhat wasteful designs and worthwhile economies may be obtained by considering interactive buckling.

For sway frames, i.e. those which resist sidesway by the bending stiffness of the frame, instability must be taken into account by using effective lengths obtained by an approved method. This will mean that almost all of the columns will have effective lengths considerably in excess of their true lengths measured from beam centre line to beam centre line.

Multistorey rigidly-jointed sway frames may be designed plastically provided that they are braced against sway out of their own planes and due allowance is made for the effects of frame instability. The approach adopted has its origin in the interaction curves governing column design. One of these is the Perry Robertson equation shown plotted in Fig. 2.16. One of the earliest formulations for this interaction was that suggested by Rankine which is of a very simple form:

$$\frac{1}{P_F} = \frac{1}{P_p} + \frac{1}{P_{cr}} \tag{4.28}$$

where P_F is the predicted failure load
        P_p is the predicted squash load
        P_{cr} is the Euler instability load.

Merchant argued that like an isolated strut a complete structure has an elastic critical load (analogous to P_{cr}) and a plastic collapse load (analogous to P_p). He suggested that the simple empirical formula proposed by Rankine might lead to quite reasonable predictions of failure loads.

Theoretical and experimental studies have shown that for bare structural frameworks Eq. 4.28 represents an approximate lower bound solution to the true failure load. In certain circumstances it may lead to slightly non-conservative solutions but it can give significantly oversafe solutions.

Work by Wood and Horne has led to this approach appearing in the new ECCS code and also in proposals for the new draft British Standard Code for Structural Steelwork in somewhat modified forms. Equation 4.28 may be written in terms of load factors as

$$\frac{1}{\lambda_F} = \frac{1}{\lambda_p} + \frac{1}{\lambda_{cr}}$$

where

λ_F is the load factor at failure

λ_p is the load factor at rigid collapse

λ_{cr} is the load factor at instability

or

$$\lambda_F = \frac{\lambda_p}{(1 + \lambda_p/\lambda_{cr})} \tag{4.29}$$

In the ECCS recommendations for a very stocky structure (defined by $\lambda_{cr} \geqslant 10\lambda_p$) then it is argued that the influence of strain hardening together with a minimum of restraint provided by cladding will result in the structure being able to carry its full rigid-plastic collapse load predicted by simple theory, i.e.

$$\lambda_F = \lambda_p \tag{4.30}$$

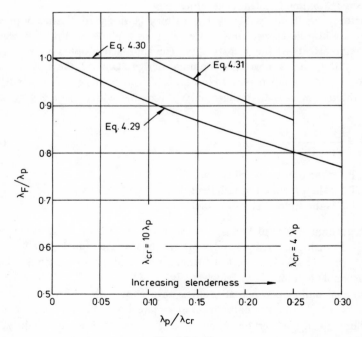

Fig. 4.44

If a structure is of intermediate slenderness (defined by $10\lambda_p > \lambda_{cr} > 4\lambda_p$) then the failure load may be obtained from Eq. 4.31 provided that certain limitations are satisfied.

$$\lambda_F = \frac{\lambda_p}{(0{\cdot}9 + \lambda_p/\lambda_{cr})} \tag{4.31}$$

This is equation 4.29 with the constant adjusted so that equations 4.30 and 4.31 are consistent at $\lambda_{cr} = 10\lambda_p$, whilst retaining the same form as the Merchant–Rankine formula. The plot of Fig. 4.44 compares these equations and also shows the limiting slendernesses.

Identical recommendations have been made for the draft of the new British Code but the equations have been recast in terms of modified design stresses for rigid-plastic analysis. Within the range of slenderness for which equation 4.31 is valid the failure load of a structure is given by the rigid plastic collapse load reduced by a factor lying between 1 and $0{\cdot}87$. Thus, it is argued, the reduction in load carrying capacity may be thought of as a reduced design strength p'_Y to be used in place of the specified material design strength p_Y. For stocky frames $\lambda_{cr} > 10\lambda_p$ the design failure load is equal to the rigid plastic collapse load and hence $p'_Y = p_Y$.

However for the intermediate slenderness range equation 4.31 must be re-examined.

Define a ratio α by

$$\alpha = \frac{\text{Elastic Critical Load}}{\text{Factored Design Load}} = \frac{\lambda_{cr}}{\lambda_F} \tag{4.32}$$

and note from the definition of the reduced design strength p'_Y that

$$\frac{\lambda_F}{\lambda_p} = \frac{p'_Y}{p_Y} \tag{4.33}$$

Now

$$\frac{\lambda_F}{\lambda_p} = \frac{\lambda_F}{\lambda_{cr}} \times \frac{\lambda_{cr}}{\lambda_p}$$

Thus

$$\frac{p'_Y}{p_Y} = \frac{1}{\alpha} \times \frac{\lambda_{cr}}{\lambda_p} \tag{4.34}$$

Also equation 4.31 may be rearranged as

$$\frac{\lambda_F}{\lambda_p} = \frac{1}{(0{\cdot}9 + \lambda_p/\lambda_{cr})}$$

and, using equations 4.33 and 4.34, this becomes

$$\frac{p'_Y}{p_Y} = \frac{1}{(0{\cdot}9 + p_Y/\alpha p'_Y)}$$

Thus

$$p'_Y = \frac{(\alpha - 1)}{0 \cdot 9\alpha} p_Y \tag{4.35}$$

It is desirable to express the limiting conditions of equation 4.35 in terms of α.

Now

$$\alpha = \lambda_{cr}/\lambda_F = (\lambda_{cr}/\lambda_p)/(\lambda_p/\lambda_F)$$

but

$$\lambda_p/\lambda_F = 1/(0 \cdot 9 + \lambda_p/\lambda_{cr})$$

Therefore

$$\alpha = (\lambda_{cr}/\lambda_p)[0 \cdot 9 + (\lambda_p/\lambda_{cr})]$$

Thus when

$$\lambda_{cr}/\lambda_p = 10, \quad \alpha = 10$$

and when

$$\lambda_{cr}/\lambda_p = 4, \quad \alpha = (4)(0 \cdot 9 + 0.25) = 4 \cdot 6$$

Thus the range of applicability of equation 4.35 becomes $10 > \lambda_{cr}/\lambda_F > 5$ if (as seems likely) $4 \cdot 6$ is rounded up to $5 \cdot 0$.

The application of the modified Merchant—Rankine formula thus becomes one of establishing the ratio α, which requires the assessment of the elastic critical load, determining whether a reduced design strength should be applied, and if so employing equation 4.35. A rigid-plastic analysis may then be carried out leading to a failure load which corresponds to that which would be predicted by a straightforward application of equation 4.31. This latter approach has the advantage in that, to allow for the effects of overall instability, it relates to a reduced design strength, which lies within the closely spaced bounds of p_Y and $0 \cdot 87 p_Y$, before a normal rigid-plastic process is pursued. Additionally it can be seen that equation 4.35 does not require a highly accurate determination of λ_{cr} and an error of 10% in λ_{cr} will result in an error of some 2% in p'_Y and hence in λ_F.

For most very slender frames (those with λ_p/λ_{cr} exceeding about $0 \cdot 25$ or α greater than $5 \cdot 0$) it is probable that such structures would suffer from excessive deformations and thus the formulae cover most of the practical range. Should such a structure be required, however, it is recommended that a more rigorous analysis be carried out in which plastic behaviour is considered together with the effects which result from changes in geometry.

4.14 Bibliography

1 Bleich, F. (1952), *Buckling Strength of Metal Structures*, New York, McGraw-Hill ch. 7.

2 Timoshenko, S. P., and Gere, J. M. (1961), *Theory of Elastic Stability*, 2nd edn., New York, McGraw-Hill, ch. 2.

3 Galambos, T. V. (1968), *Structural Members and Frames*, Englewood Cliffs, N.J., Prentice-Hall, chs. 4 and 6.

4 Johnston, B. G. (ed.) (1977), *Guide to Design Criteria for Metal Compression Members*, 3rd edn., SSRC, New York, Wiley, ch. 15.

5 Trahair, N. S. (1977), *The Behaviour and Design of Steel Structures*, London, Chapman and Hall, ch. 7.

6 Winter, G. et al. (1948), *Buckling of Trusses and Rigid Frames*, Bulletin 36, Ithaca, U.S.A., Cornell University.

7 Merchant, W. (1954), 'The Failure Load of Rigid Jointed Frameworks as Influenced by Stability', *J. Inst. Struct. E*.

8 Bolton, A. (1955), 'A New Approach to the Critical Load of Rigidly Jointed Trusses', *J.Inst. Struct.E*.

9. Bolton, A. (1955), 'The Critical Load of Portal Frames when Sidesway is Permitted', *J.Inst. Struct.E*.

10. Merchant, W., et al. (1955–6), 'Critical Loads of Tall Building Frames', *J.Inst. Struct.E*.

11. Wood, R. H. (1974), 'Effective Lengths of Columns in Multistorey Buildings', *J.Inst. Struct.E*.

12. Horne, M. R. (1975), 'An Approximate Method for Calculating the Elastic Critical Loads of Multistorey Plane Frames', *J.Inst. Struct.E*.

13. Livesley, R. K., and Chandler, D. B. (1956), *Stability Functions for Structural Frameworks*, Manchester University Press.

14. British Standard 449, Part 2: (1969), *Specification for the Use of Structural Steel in Building*, London, BSI.

15. *Draft Standard Specification for the Structural Use of Steelwork in Building, Part 1: Simple Construction and Continuous Construction* (1977), London, BSI.

16. *Recommendations for Steel Constructions* (1976), Rotterdam, European Convention for Constructional Steelwork.

Selected tabulated values of stability functions

Table 1

Values of s, c, $s(1 - c^2)$ and $(sc)^2$ for positive values of ρ i.e. compression

ρ	s	c	$s(1 - c^2)$	$(sc)^2$
0·00	4·000	0·500	3·000	4·000
0·04	3·947	0·510	2·920	4·053
0·08	3·894	0·521	2·838	4·109
0·12	3·840	0·532	2·755	4·166
0·16	3·785	0·543	2·669	4·224
0·20	3·730	0·555	2·581	4·285
0·24	3·674	0·568	2·490	4·348
0·28	3·617	0·581	2·397	4·413
0·32	3·560	0·595	2·302	4·480
0·36	3·502	0·609	2·204	4·549
0·40	3·444	0·624	2·102	4·621
0·44	3·385	0·640	1·997	4·695
0·48	3·325	0·657	1·889	4·773
0·52	3·264	0·675	1·777	4·852
0·56	3·203	0·694	1·662	4·935
0·60	3·140	0·714	1·541	5·021
0·64	3·077	0·735	1·417	5·110
0·68	3·013	0·757	1·287	5·202
0·72	2·948	0·781	1·151	5·299
0·76	2·883	0·806	1·010	5·398
0·80	2·816	0·833	0·862	5·502
0·84	2·748	0·862	0·707	5·610
0·88	2·680	0·893	0·544	5·722
0·92	2·610	0·926	0·373	5·839
0·96	2·539	0·962	0·192	5·961
1·00	2·467	1·000	−0·000	6·088
1·04	2·394	1·042	−0·204	6·221
1·08	2·320	1·087	−0·420	6·359
1·12	2·245	1·136	−0·652	6·503
1·16	2·168	1·190	−0·901	6·654

Table 1 (continued)

ρ	s	c	$s(1 - c^2)$	$(sc)^2$
1·20	2·090	1·249	−1·169	6·812
1·24	2·011	1·314	−1·459	6·977
1·28	1·930	1·386	−1·775	7·150
1·32	1·848	1·465	−2·120	7·331
1·36	1·764	1·555	−2·501	7·521
1·40	1·678	1·656	−2·922	7·720
1·44	1·591	1·770	−3·393	7·930
1·48	1·502	1·900	−3·923	8·150
1·52	1·411	2·051	−4·527	8·381
1·56	1·319	2·227	−5·222	8·625
1·60	1·224	2·435	−6·032	8·881
1·64	1·127	2·684	−6·992	9·152
1·68	1·028	2·988	−8·150	9·438
1·72	0·927	3·367	−9·580	9·739
1·76	0·823	3·852	−11·395	10·059
1·80	0·717	4·497	−13·783	10·397
1·84	0·608	5·393	−17·078	10·755
1·88	0·496	6·722	−21·935	11·135
1·92	0·382	8·899	−29·847	11·538
1·96	0·264	13·109	−45·084	11·967
2·00	0·143	24·684	−86·864	12·424
2·04	0·018	197·386	−709·240	12·911
2·08	−0·110	−33·292	121·901	13·431
2·12	−0·242	−15·436	57·487	13·987
2·16	−0·379	−10·085	38·132	14·582
2·20	−0·519	−7·511	28·781	15·219
2·24	−0·665	−5·998	23·254	15·904
2·28	−0·815	−5·003	19·592	16·640
2·32	−0·971	−4·299	16·977	17·433
2·36	−1·133	−3·775	15·011	18·288
2·40	−1·301	−3·370	13·472	19·213
2·44	−1·475	−3·048	12·231	20·215
2·48	−1·656	−2·787	11·205	21·302
2·52	−1·845	−2·570	10·339	22·484
2·56	−2·043	−2·387	9·595	23·773
2·60	−2·249	−2·231	8·948	25·181
2·64	−2·465	−2·097	8·376	26·723
2·68	−2·692	−1·981	7·866	28·417
2·72	−2·930	−1·878	7·407	30·281
2·76	−3·180	−1·788	6·989	32·341
2·80	−3·445	−1·708	6·606	34·623
2·84	−3·725	−1·637	6·252	37·160
2·88	−4·021	−1·573	5·923	39·990
2·92	−4·337	−1·515	5·616	43·159
2·96	−4·673	−1·463	5·326	46·722

Table 1 (continued)

ρ	s	c	$s(1-c^2)$	$(sc)^2$
3·00	−5·032	−1·416	5·053	50·746
3·04	−5·417	−1·373	4·793	55·312
3·08	−5·832	−1·334	4·544	60·519
3·12	−6·281	−1·298	4·306	66·491
3·16	−6·767	−1·266	4·077	73·381
3·20	−7·297	−1·236	3·856	81·383
3·24	−7·878	−1·209	3·641	90·744
3·28	−8·518	−1·184	3·432	101·783
3·32	−9·227	−1·162	3·228	114·917
3·36	−10·018	−1·141	3·028	130·700

Table 2

Values of s, c, $s(1 - c^2)$ and $(sc)^2$ for negative values of ρ i.e. tension

ρ	s	c	$s(1 - c^2)$	$(sc)^2$
0·00	4·000	0·500	3·000	4·000
−0·20	4·257	0·455	3·374	3·756
−0·40	4·501	0·418	3·714	3·545
−0·60	4·735	0·387	4·025	3·362
−0·80	4·959	0·361	4·314	3·202
−1·00	5·175	0·338	4·583	3·060
−1·20	5·382	0·318	4·837	2·935
−1·40	5·583	0·301	5·077	2·824
−1·60	5·777	0·286	5·305	2·724
−1·80	5·965	0·272	5·523	2·635
−2·00	6·147	0·260	5·731	2·554
−2·20	6·324	0·249	5·932	2·481
−2·40	6·496	0·239	6·125	2·414
−2·60	6·664	0·230	6·311	2·354
−2·80	6·828	0·222	6·491	2·298
−3·00	6·988	0·215	6·666	2·247
−3·20	7·144	0·208	6·836	2·200
−3·40	7·297	0·201	7·001	2·157
−3·60	7·446	0·195	7·162	2·117
−3·80	7·593	0·190	7·319	2·080

Table 3

Values of m, n and o for positive values of ρ i.e. compression

ρ	m	n	o
0·00	1·000	1·000	1·000
0·02	1·017	0·933	1·034
0·04	1·034	0·865	1·069
0·06	1·052	0·794	1·106
0·08	1·071	0·722	1·145
0·10	1·091	0·647	1·186
0·12	1·112	0·570	1·229
0·14	1·134	0·491	1·274
0·16	1·156	0·408	1·321
0·18	1·180	0·323	1·372
0·20	1·205	0·235	1·425
0·22	1·231	0·144	1·481
0·24	1·259	0·049	1·540
0·26	1·288	−0·050	1·603
0·28	1·319	−0·153	1·669
0·30	1·351	−0·260	1·740
0·32	1·385	−0·372	1·816
0·34	1·422	−0·489	1·896
0·36	1·460	−0·612	1·982
0·38	1·501	−0·742	2·074
0·40	1·545	−0·878	2·172
0·42	1·592	−1·022	2·278
0·44	1·642	−1·174	2·392
0·46	1·696	−1·336	2·515
0·48	1·754	−1·508	2·648
0·50	1·817	−1·691	2·792
0·52	1·885	−1·887	2·949
0·54	1·958	−2·099	3·120
0·56	2·039	−2·326	3·307
0·58	2·127	−2·573	3·514
0·60	2·223	−2·842	3·741
0·62	2·330	−3·136	3·994
0·64	2·449	−3·459	4·276
0·66	2·582	−3·817	4·592
0·68	2·731	−4·216	4·949
0·70	2·900	−4·665	5·354
0·72	3·094	−5·173	5·819
0·74	3·317	−5·754	6·357
0·76	3·577	−6·427	6·986
0·78	3·884	−7·217	7·732
0·80	4·253	−8·159	8·629
0·82	4·703	−9·303	9·728

Table 3 (continued)

ρ	m	n	o
0·84	5·266	−10·725	11·105
0·86	5·990	−12·544	12·878
0·88	6·956	−14·959	15·247
0·90	8·307	−18·326	18·567
0·92	10·334	−23·361	23·554
0·94	13·711	−31·728	31·874
0·96	20·466	−48·430	48·527
0·98	40·731	−98·465	98·514
1·00	∞	∞	∞
1·02	−40·326	101·464	−101·514
1·04	−20·061	51·429	−51·528
1·06	−13·306	34·726	−34·876
1·08	−9·928	26·356	−26·558
1·10	−7·902	21·319	−21·573
1·12	−6·550	17·949	−18·254
1·14	−5·585	15·531	−15·889
1·16	−4·861	13·707	−14·119
1·18	−4·298	12·281	− 12·746
1·20	−3·847	11·131	−11·651
1·22	−3·478	10·183	−10·758
1·24	−3·171	9·387	−10·018
1·26	−2·911	8·707	−9·394
1·28	−2·688	8·117	−8·861
1·30	−2·495	7·601	−8·403
1·32	−2·326	7·144	−8·004
1·34	−2·176	6·736	−7·655
1·36	−2·043	6·368	−7·346
1·38	−1·925	6·034	−7·073
1·40	−1·818	5·729	−6·829
1·42	−1·721	5·448	−6·610
1·44	−1·633	5·189	−6·414
1·46	−1·552	4·948	−6·236
1·48	−1·479	4·723	−6·076

Table 4

Values of m, n and o for negative values of ρ i.e. tension

ρ	m	n	o
0·00	1·000	1·000	1·000
−0·20	0·863	1·585	0·734
−0·40	0·764	2·063	0·555
−0·60	0·689	2·471	0·430
−0·80	0·631	2·830	0·340
−1·00	0·584	3·153	0·272
−1·20	0·545	3·449	0·221
−1·40	0·513	3·722	0·181
−1·60	0·485	3·977	0·149
−1·80	0·461	4·217	0·125
−2·00	0·440	4·444	0·105
−2·20	0·421	4·661	0·088
−2·40	0·405	4·868	0·075
−2·60	0·390	5·066	0·064
−2·80	0·377	5·257	0·055
−3·00	0·364	5·442	0·047
−3·20	0·353	5·620	0·041
−3·40	0·343	5·793	0·035
−3·60	0·334	5·961	0·031
−3·80	0·325	6·124	0·027

Index